Y0-DLJ-574

~Green Beings Save Genes~

The Complete Guide to

Energy

Conservation

For

Smarties

Written by Susan Hartsfield, MSN, NP
Illustrated by Rajeev Athale

Copyright © 2008 by Susan Hartsfield
First Edition
All rights reserved.

Published by:
Green Being Publishing Co.
Printed and Bound by United Book Press
www.unitedbookpress.com
Baltimore, MD

Library of Congress Control Number: 2008906659

Green Being ©
Publishing Company
Annapolis Maryland
2008

Acknowledgements

This book was the idea of my most unique friend, Jim Reardon of Perpetual Prosperity Pumps Foundation. Jim's altruism and ambition to improve the lives of people around the world inspired me to join in another great challenge facing us today - Global Warming. He is a true role model, spending his time and energy to bring the people of Ghana out of poverty, one tennis shoe at a time. Please visit his web site and learn how you can help, too. It is gratifying to know that donating your old tennis shoes help to provide the basic substance of life – food and water to so many! http://www.pppafrica.org

This book would be rather tedious without the detailed and graphic cartoon commentaries of Rajeev Athale. It is a wonder he tolerated all my requests for more and more. Thank you, Rajeev. And to my wonderful environmental friend and Editor Cherie Yelton, I owe extreme gratitude. Also, special thanks are sent to Lisa Wallace for her clever logo design.

The book would have been impossible without the piles of research and data from so many environmental groups and government agencies that edify contribute to and rouse ordinary citizens to save and savor this beautiful planet on which we live.
And the most difficult part of writing a book must be the computer graphic design and layout. I would have been lost or at least delayed by months or years had it not

been for the help of Jack Hartsfield. There isn't a thank you big enough for his effort.

And finally I send an overwhelming thank you to each and everyone sharing the challenges of energy efficiency. Taken seriously, this worthwhile cause will add pleasure and purpose to your life.

About the Author:

Susan Hartsfield's career as a Nurse Practitioner has spanned nearly three decades of providing medical care for hospitalized patients, and education for nurses and resident physicians. At the core of her beliefs is a notion that personal responsibility is paramount to maintain one's health as well as the health of the planet.

What has that got to do with the campaign toward public awareness and the effort to decrease global warming, you may ask? The two are closely tied by the apparent needs of the public to understand *why* they are asked to change—whether it be diet and exercise or turning off the lights in an unused room or the use of a canvas grocery bag. The question *why*, must be understood.

Her love of nature keeps Susan busy kayaking and biking or gardening where she can observe and cherish the great outdoors. Combine that with a lifelong habit of thrift learned from parents who survived the Great Depression, and the stage was set for writing Energy Conservation for Smarties.

Susan has used her "bedside manner" with patients— simple explanations, empathy, non-judgmental acceptance and humor to communicate the needs for us to make informed choices to reduce energy consumption and she shows us how to go about doing it to save money, save the planet and save humanity! Susan holds a Masters Degree in Nursing and dual certifications as an Acute Care and Adult Care Nurse Practitioner

This book is dedicated to my children

Melissa and Sarah

And to all the children of the world

About the Illustrator:

About me ....
My name is Rajeev Athale and I hail from India; the ancient land of snake charmers and computer programmers.

I have been drawing since I was a kid, can't even remember when I started. My first cartoon was published in a magazine when I was nineteen.
In my other life, I am an architect and run my tiny design firm. I live in Indore in central India with my wife and my seven-year-old daughter.
I hope that you liked my cartoons. Please keep visiting CartoonHub.com as new cartoons are added frequently.

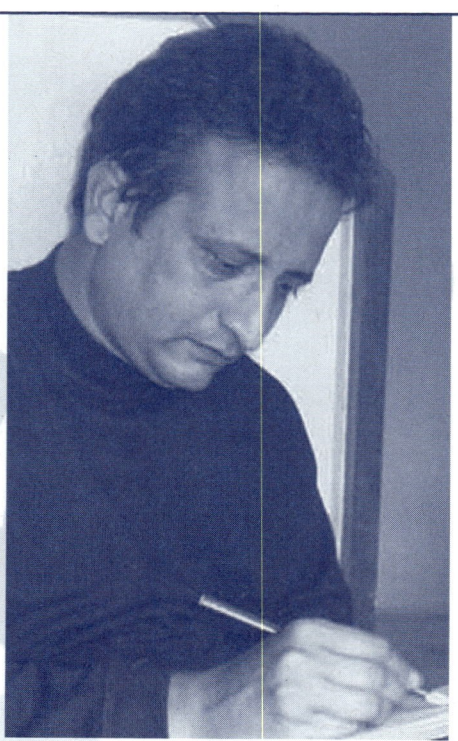

~ Table of Contents ~
100+ FREE Ideas
Curbing Emissions Saves You $$$
Resources and Links

	Page
Introduction: In a Nutshell	14

Phase One

Chapter One: Efficiency	23
Chapter Two: Flips and Turns	30
Chapter Three: Driving Cents to the Bank	51
Chapter Four: Dialing for Dollars	69
Chapter Five: Garden Delight	76
Chapter Six: Wee-Bit of Will Power	91
Chapter Seven: Reduce Reuse Recycle	108
Chapter: Eight Good Ole Uncle Sam	133

Real Action
Small Investments Big Savings
Resources and Links

Phase Two

	Page
Chapter Nine: Abe to Ulysses	141
Chapter Ten: Ulysses to Ben	165
Chapter Eleven: As Much as You'd Like	173
Chapter Twelve: Premeditated Assault on Carbon	201

Smart Growth
Big Investments Lifetime Savings
Resources and Links

Phase Three

	Page
Chapter Thirteen: Empowering your Home	214
Chapter Fourteen: Empowering your Transportation	235
Chapter Fifteen: The Second Biggest Investment	243
Chapter Sixteen: The Abode	257
Chapter Seventeen: The Odds for the End	266
Chapter Eighteen: Conclusion	272

Finale

Resource Guide	277
Glossary	284
Index	294
Global Baskets	297
Recycled Book Paper	300

Forward:

This book is a collection of knowledge and wisdom studied and discovered by dozens of environmental groups, researchers, scientists and government agencies. These tactics have been complied in a practical and organized manner for the reader to progress through at an easy light-hearted pace.

There are many things that each of us can do to conserve energy. You have the backbone for action and the brains for joining in the most worthwhile purpose for improving our planet. I have tried to lay the book out in a clear and organized fashion to focus your transition to a greener world in phases from easiest and free in Phase I, to moderate costs and efforts in Phase II, and most expensive and extensive suggestions in Phase III

Let's use toilets and water conservations as an example:

- In Phase One—remember, everything is free here--to save water and money, just add a brick to the holding tank on the back of the "john".
- In Phase Two, for a small cost and little bit more effort, you can purchase a new low flow toilet.
- In Phase Three, if you're in the process of remodeling or building, you'll find resources on waterless, composting toilets.

The point is, depending on circumstances; every person on the earth can contribute to efficiency.

This book serves as a reference point for computer web sites and links, telephone numbers and specific companies or establishments to get you started on every task.

Introduction:

A haunting from the past

> "There is no energy policy that we can develop that would do more good than voluntary conservation. There is no economic policy that will do as much as shared faith in hard work, efficiency, and in the future of our system. There is no way that I, or anyone else in the Government, can solve our energy problems if you are not willing to help. I know that we can meet this energy challenge if the burden is borne fairly among all our people--and if we realize that in order to solve our energy problems we need not sacrifice the quality of our lives."
>
> President Jimmy Carter 1977

Though this book is directed toward the American reader, and speaks of air pollution, acidic oceans, melting ice caps and rising sea levels, it must be remembered that these problems and water and food shortages are not a local problem, or even a state or a federal problem. They are global problems. That's one reason why they call it Global Warming! We are all in this together.

It is this *collective-will* that brought the United Nations together for a conference in 1997. At this first

Convention on Climate Change, the Kyoto Protocol was conceived. This treaty attempts to share human destiny, equally by all countries, to reduce emission of carbon dioxide and other greenhouse gasses (GHG) below the 1990 levels prior to 2012. Before explaining the Kyoto Treaty, one must understand some important things about carbon emissions and reducing them to the specified target level.

How much actual carbon are we talking about, you may wonder? Well, like all new fields of study, the information changes as we learn more. At the beginning of the industrialization the US concentration of carbon dioxide (CO_2) was 275 parts per million (ppm) in the atmosphere. But that number has already grown to its current status of 383 ppm.

For the last thirty years, scientists have been studying the climate and the glaciers trying to estimate just how much carbon pollution is acceptable without warming and harming the planet. They believed the amount was 550 parts per million.

Roughly on that same timeline, physicians were studying cholesterol. It was thought that a cholesterol level of 220 + the age was acceptable. Wow, we now know that all that cholesterol fat is causing death and tragedy in human life. Today we know the LDL cholesterol should be well below 100 and probably closer to 60. That's what's

happening in the climatologist and glaciologist's world too. They now know that 550 ppm were way too high.
Evidence for this was when an alarming incident occurred and almost no one noticed. In March of 2002, Larsen B, a massive ice shelf off the Antarctic Peninsula, stable for more than 10,000 years suddenly cracked, broke off and disintegrated into the ocean. At the same time a giant Greenland ice sheet was accelerating its melting just like the West Antarctic ice shelf. Ice is still melting and, as it does - it settles in small glacial cracks. When the melt water freezes again the next season, it will expand farther and cause even deeper cracks. This arctic cycle continues season after season. Each melt-and-freeze cycle causes deeper and more prominent cracks until the inevitable happens—disintegration of the entire shelf occurs as it did with Larsen B.

Scientists now know the warming planet is due to increasing levels of carbon dioxide. Warming is happening quickest at the poles.

This is not a scientific book - but it is important that you understand why those numbers and this process are so important to you. Hang on here just a little more.
For every 150 cubic miles of ice that melts into the sea, the global sea level will raise $1/16^{th}$ of an inch. That sounds ok to me. But considering the ice shelves are over two miles deep, it is hard to realize the amount of water that translates into. For instance if the Western Arctic shelf melts, the sea level would raise 19 feet. If all

three of the ice shelves melt: the Western, the Greenland and the Larsen B, the ocean around the entire world will rise about 213 feet. The Statue of Liberty will be 50 foot underwater! If you are among the 33 percent of people that currently live at sea level to 300 foot above sea level, you may lose your home and even worse, your life. http://www.sciam.com/article.cfm?id=the-unquiet-ice

NASA scientist Dr. James Hansen has made it clear that the planet cannot sustain the glaciers and climate we know and enjoy if the CO_2 is greater than 350 ppm! And this is the worrisome part because at 383 ppm we are already 33 parts per million over that tipping point. One Earth equals 350. Know that number; become familiar with its meaning. (This book was priced to bring those numbers home, 1 Earth, 350, hence $13.50). Thank heavens; they were wrong about the 550 or this book would cost two dollars more. Let's not be focused on a certain percentage of CO_2 being lowered by any given year, the bottom line is that we must reach 350 ppm now.

So, back to the Kyoto treaty, by November of 2007, 174 countries had accepted this moral responsibility and ratified the Kyoto treaty. The United States (US) has still not signed this agreement. The current administration declined to ratify Kyoto claiming it would hurt US companies' competitiveness because developing countries like China and India weren't required to curb emissions. However, even though China as a whole has surpassed the US in carbon emissions (Hong Kong ratified the Kyoto treaty in 2003), here are some of the things they *have* accomplished; see http://www.worldchanging.com/archives/004573.html for more.

1. In January 2008 China banned all free plastic shopping bags.
2. The Chinese still use bicycles for 40 percent of their transportation. There are over five million bicycles, but only one car for every tenth person.
3. All government employees must take public transportation or ride a bike to work.

India too has completed their own strategic achievements in the effort to decrease carbon emissions besides producing the Nano, the most affordable energy efficient auto on the planet.

Yet, in spite of India's meteoric growth in the recent past, an American produces 16 times more greenhouse gas emissions than the average Indian and the Indian government points out that it contributes only 4.6 percent of the world's GHG although its people represent 17 percent of the world's population according to *the New York Times*

Let us not fret about who is doing what. We all need to put our heads and hearts together on this one. We must decrease carbon emissions below 350 ppm. Remember, think globally, and act locally.

We the people of the US can show the United Nations and the world just what we are capable of creating and accomplishing when challenged. When we share ideas and efforts, network our talents, and motivate one another the sky is the limit. In fact if every household in the US made <u>energy-efficient</u> choices, we could reduce our emissions by up to two-thirds annually. We could save 800 million tons of global warming pollutants -- this is more than the emissions released by over 100 countries!

I admit this stuff can be confusing. Here is an urban legend traveling across the web. It is hard to determine what is true and what is not true.
Example: Choosing Fuel Wisely
Companies claimed to import Middle Eastern oil:
- Shell   205,742,000 barrels
- Chevron/Texaco   144,332,000 barrels
- Exxon/Mobil   130,082,000 barrels
- Marathon/Speedway 117,740,000 barrels
- Amoco   62,231,000 barrels

Citgo gas is imported from South America, from a Dictator who hates Americans.
Here are some large companies that are claimed not to import oil from Middle Eastern refineries:
- Sunoco   0 barrels
- Conoco   0 barrels
- Sinclair   0 barrels

- BP/Phillips   0 barrels
- Hess.   0 barrels
- ARCO   0 barrels

Is it true? I don't know. But this is true and was reported by David Ivanovich on May 2, 2008 in the *Houston Chronicle*: "The (petroleum) reserve now contains more than 701 million barrels of oil, up 27 percent since Bush took office. The administration is injecting 67,000 to 68,000 barrels a day to the reserve and over the next three months will add nearly 8 million new barrels". Here is another shocking truth. The Washington Post reported on 4/30/08 "Two oil giants, BP and Royal Dutch Shell, announced record windfall profits yesterday totaling $17 BILLION dollars in the first three months of the year". And Exxon trailed behind with only a net income of only $10.9 BILLION in profits. They have slipped a little from their $11.7 BILLION earned in the last quarter of 2007. But overall they can't complain with an annual profit of $40.6 BILLION DOLLARS in 2007! http://www.nytimes.com/2008/05/02/business/02oil.html?ref=business

There is something drastically wrong with that picture. At the moment it does **not** appear there is a 'supply' problem with over 700 million barrels in reserve. Maybe it is simply a greed and profit scam at our expense? If you'd rather deposit your hard earned money in your own bank account instead of theirs, this book is for you.

And what is the likely solution to not buying imported oil? Off shore drilling in the US, of course...in the Rocky Mountains, in the Arctic National Wildlife Refuge, in the Gulf of Mexico? That's a SOLUTION?? It's a no-brainer...there *are* **no** good sources for oil! The answer...**conservation**, and **renewable energy** and what this book is all about: **Efficiency**. These are the only solutions to cleaning up the air we breathe, the lands and waters where we get our food, and leveling the social playing field.

Think globally and act locally

Think globally and act locally

Think globally and act locally

PHASE ONE
Chapter One

Eeeeeeeezy!

What is this Energy Conservation for Smarties all about?

At last a book that acknowledges your natural intellect has been written. The book for Smarties is here. Just picking it up places you way ahead of those less inclined to make change happen.
You may have been feeling a bit perplexed with those lists all over town giving you so many disorganized goals, you've decided to do your own investigating. You want to know everything about the easiest and least expensive energy tactics? Well, they are right here endorsed in Smartie Handbook! It is an encyclopedia of tactics to decrease expensive electricity! It encourages a little extra effort from everybody, everyday to end emissions and enhance the Earth's environment to be everlasting. It contains endless opportunities for enlarging individual economic$ on every page! It is an energy revolution that will empower you. And it is FREE. It is called...

EFFICIENCY!

~GREEN BEINGS –$AVE GENES~

More is better?

There is an entrenched idea floating around out there in the minds of many:"More is better". That is just n*ot true*! More consumption has a tipping point, too. To compare to the health field more food, more tobacco, more alcohol, more recreational drugs, more sedentary, more entertainment, more partners, more stress, and yes even more pharmaceutical drugs lead to "*too much of a good thing*".

The slogan seems to violate our way of thinking. But, yes, *all good things do come to an end* and each can eventually lead to disease evidenced by fever, sweats, chills, dehydration, swelling, vomiting, tumors and so on.

This notion holds true for an even a larger living specimen: the Earth which also is showing signs of dis~ease. More air pollution, more carbon emissions, more mountain tops removed, more mining, more water pollution, more fertilizer, more pesticides, more landfills, more toxins, loss of species, loss of rainforest, loss of wet lands, coral reefs and more. It's all evidenced by rising temperatures, raging fires, melting ice caps, droughts, floods, hurricanes, tsunamis, tornadoes, quakes, algae blooms and so on. WHEW!

Cancer brews in an unsuspecting host for years and, when it finally declares itself, the swiftness with which it destroys can completely overwhelm the individual. Just as with cancer, scientists had under-estimated just how quickly the Earth's tipping point actually occurs once it is started in motion.

To limit these disastrous events from happening, preventive medicine can be promoted through a healthy lifestyle, diet and exercise. On a much larger scale, to prevent further disastrous changes in the Earth's atmosphere and deteriorating climate we all need to follow a similar path. Remember, it is called

EFFICIENCY

It is really as simple as being aware that our behaviors and habits that extend beyond our own bodies can be very harmful to the environment. Most people have heard of a healthy lifestyle to promote longevity and quality life.

Now it is time to look beyond the personal diet and exercise level and raise the bar of recognition to include our global responsibility. As stewards of the Earth and with collective actions our common agenda will focus on our commitment to reduce, reuse and recycle to succeed in this battle.

It is no longer a question of whether we should worry about a healthy planet for our children and grandchildren. The question is: *how* do we reduce our pollution below the tipping point of 350 ppm of carbon emissions?
The U.N. Intergovernmental Panel on Climate Change (IPCC) released a report on November 29, 2007, summarizing the findings of the panel's 2,500 scientists. This is their verdict as described by the Chairman of the IPCC:

"If there's no action before 2012, that's too late. What we do in the next two to three years will determine our future. This is the defining moment."

With that said let's join in the educational excursion into the world of efficiency.

Energy Usage in the Home

Electricity use in the typical home is about 877 kilowatts per month and 26,730 pounds of carbon emission into the atmosphere each year. That is just about 75 pounds of carbon per day. An additional 25-45 pounds of carbon are emitted each day from individual private transportation.

According to Rocky Mountain Institute at http://www.newdream.org/ consumer/PKG22.pdf, home energy use is divided among five major areas:

Heating and cooling	60 percent
Water Heater	16 percent
Refrigerator	12 percent
Lighting	7 percent
Computers, electronics	5 percent

To prevent further rise of green house gas (GHG) in the atmosphere without causing a *climate snap (tipping point)*, it is believed by authorities that Americans need to reduce their energy use to less than ten pounds of carbon emissions per day.

You can do it and save money! Forget Geico! This book is guaranteed to save you thousands of dollars through efficiency awareness and conservation while lowering your level of GHG emissions. The first step is to read your current utility bill. Pull out an old bill. Use it as a bookmark so you will have it handy to compare your savings as your travel through the phases of energy conservation outlined in this book.

Fill in the information you learn from your utility bill right here. You will enjoy looking back at this page for your ever-increasing savings.

~GREEN BEINGS –SAVE GENES~

◆ Month:

◆ Utility:

◆ Costs:

◆ Kilowatts Used:

Your mission, should you decide to accept it, is first to reduce major expenses. This will not only help the planet but you will see your monthly expenses decrease and your savings increase. Month to month, as you follow the guidelines you will watch your kilowatts decrease from average 900 to less than 400 kilowatts each month. According to the Alliance to Save Energy in Washington DC, the average American household spends about $2200 each year on energy bills. This figure is guaranteed to continue rising year after year. So get ready to climb the rungs of opportunity and laugh all the way to the bank.

Kinds of Energy We Use in Our Home

Water was the bargain: in many places water was plentiful until severe droughts and population demands surpassed availability. The average price of water in the US is about a dollar and a half for 1,000 gallons. At that price, a gallon of water costs less than one penny. But prices are rising and restrictions are becoming a way of life for many. Today there are severe shortages in several areas of this country. Much of the water we use is 'warm or hot' so it is important to consider efficiency even where water is abundant.

Another melted glacier leaves babes stranded

Electricity is the expense: In 2007 the average costs were 10.56 cents per kilowatt (kW). Kentucky remains the lowest and Hawaii the highest costing more than 15 cents a kilowatt. US residential electricity consumption is growing at a typical rate of 2.2 percent in 2008. These official energy statistics from the US government can be found at
http://www.eia.doe.gov/emeu/steo/pub/contents.html

Home Heating Oil: The average cost in 2007 was $2.40 per gallon. But as this book is written the cost has surpassed $3.90 a gallon in a matter of months. Over 70 percent of heating oil is used in the North East.

Chapter Two

Flips and Turns

FOR EVERY SEASON,
TURN, TURN, TURN. ♪
THERE IS A REASON,
TURN, TURN, TURN ♪

See what turning dials, switches and controls do for your kilowatt usage. You will also need to ignore the comments from the pea brains making fun of you.

Turn off the lights: The first tip is the easiest and most important habit to develop. It also requires the fewest brain cells and can be achieved instantly. Poof! Five to ten percent of your electricity bill will disappear. That's how much just turning off the lights will save. Yep, you can really make a big difference with this simple step. This applies to both incandescent and fluorescent lighting.

Open wide: No, not your mouth-- the curtains. That's right: open the curtains to sit by the window and read. This is almost as easy as tip one but does require two steps. Use natural light when available.

You see, only ten to fifteen percent of the incandescent bulb provides light? The remaining 85 percent is heat. Turn off each light even though you know you will be back. But for fluorescent bulbs, the US Department of Energy (DOE) says the rule of thumb is to turn it off only if you leave the room for more than 15 minutes.
http://www.eere.energy.gov/ consumer/ your home/lighting_daylighting/index.cfm/mytopic=12280
http://www.eere.energy.gov/consumer/your_home/lighting_daylighting/index.cfm/mytopic=12280

Turn down the thermostat: In cooler areas turn the temperature down to 65 degrees Fahrenheit (We'll be talking in Fahrenheit for the rest of this book) during the day when you're home and 55 degrees at night and during days when no one is home. Make a habit of wearing sweaters or thermal undershirts. For every one-degree you turn your home's thermostat down you will save three percent of your energy bill. By lowering your thermostat five degrees at night and ten degrees during the day when you're away you can cut as much as 20 percent of

~GREEN BEINGS —SAVE GENES~   Page 31

your heating costs. Carbon savings are about 350 pounds per year for every two-degree reduction.
According to Edison Electric Institute, by lowering the heat five degrees for eight hours at night and ten degrees for eight hours during the day you will save about 3,150 pounds of GHG emissions annually.
http://www.eei.org/industry_issues/retail_services_and_delivery/wise_energy_use/100Ways.pdf
Environmental Defense Fund: US Department of Energy: http://www1.eere.energy.gov/consumer/tips/save_energy.html
http://www.eei.org/industry_issues/retail_services_and_delivery/wise_energy_use/100Ways.pdf
Environmental Defense Fund:
http://www.fightglobalwarming.com/documents/5120_BrochureR4.pdf
US Department of Energy:
http://www1.eere.energy.gov/consumer/tips/save_energy.html

Turn up the thermostat: In summer if you must use central air conditioning set the temperature to 78 degrees. For each degree of a higher setting you can expect your cooling costs to decrease by about six to eight percent. For every three degrees you increase the temperature will save 378 pounds of CO_2 per year. Amaaaazing, isn't it?

Turn it even higher or off when away. It is a myth that lowering the temperature at the onset will cool the place quicker. It will not. It simply will cost more by requiring more energy. Dress in light clothing and use ceiling fans in the rooms where you spend the most time. US Department of Energy: http://www1.eere.energy.gov/consumer/tips/save_energy.html

Turn off the ceiling fan: Remember to turn the fan off along with the lights in the summer when you leave the room. A ceiling fan cools you — not the room. In winter, reverse the paddles to force warm air downward by flipping a small switch at the top of the fan motor housing.

Now turn down the thermostat again: Turn down the thermostat on the water heater. That means lowering the temperature because the factory setting is usually set at the highest level or 145 degrees. A setting of 115 degrees provides comfortable hot water for most uses and will prevent scalding that can burn tender babies and elderly frail skin. Did you know for each ten-degree reduction in the water temperature setting a reduction of three to five percent will be reflected in your energy bill? This will save 733 pounds of carbon pollution each year. US Department of Energy:

http://www1.eere.energy.gov/consumer/tips/pdfs/energy_savers.pdf

Up, down all around...are you spinning yet? Keep going!

You mean such a little tiny action can really make a great big change???

Turn the thermostat in the refrigerator: Refrigerators are the third largest energy users in the home. Use a thermometer to set your refrigerator temperature as close to 37 degrees and your freezer as close to three degrees as possible.

To check the refrigerator temperature, place an appliance thermometer in a glass of water in the center of the refrigerator. Read it after 24 hours. To check the freezer temperature, place a thermometer between frozen packages. Read it after 24 hours. Make sure that the energy saver switch is turned on.

Also, check the gaskets around your refrigerator and freezer doors to make sure they are clean and sealed tightly. You can check this by making sure that a dollar bill closed in between the door gaskets is difficult to pull out. If it slides easily between the gaskets, replace them. Check the eHow web site for the moderately challenging task of replacement.
http://www.ehow.com/how_8259_replace-gasket-refrigerator.html
Natural Resource Defense Council:
http://www.nrdc.org/air/energy/genergy/easy.asp
Pace Law School Energy Project:
http://powerscorecard.org/reduce_energy.cfm

Turn off the dishwasher when it reaches the heat cycle: Air-dry or (bamboo) towel-dry the dishes instead of having the dishwasher pump hot air over them. Either choice is good. Use the no heat drying option if you have one or just turn the dishwasher off at the end of the rinse cycle. Here are some other Earth-saving tips about dishwasher usage:

- Run only full loads in your dishwasher. They use an incredible 860 kWh of energy. Use short cycles for all but the dirtiest dishes.

- Do not use dish soap containing phosphates. They are killing the fish. (There will be more on this topic in the near future).

- Fill the dish racks according to the manufacturer's instructions.

- Run the dishwasher at off-peak times (such as bedtime) to take advantage of lower rates offered by some electricity companies. This is more important than you think. Utilities companies build more power plants based on energy needs at peak times. When you keep your use to off peak times, you are actually helping to prevent more polluting power plant production. P-U-!

- Hand-washing vs. machine-washing, it's nearly a wash, as the saying goes. The average amount of water used to hand wash dishes is approximately nine to twenty gallons versus nine to twelve gallons used by the machine. See below.

From Natural Resources Defense Council:
http://www.nrdc.org/air/energy/genergy/easy.asp
US Department of Energy:
http://www1.eere.energy.gov/consumer/tips/save_energy.html

Turn off the water: I call this "the old-fashioned way to wash dishes." If you hand-wash dishes, and are lucky enough to have two sinks, fill one with soapy washing water and the other with rinse water. Otherwise, fill a separate basin with rinse water. This helps to trim down the average nine to twenty gallons of water used for hand washing.

Turn the washing machine to cold wash/cold rinse cycle: Rinse water should always be cold since the temperature does not affect cleaning and you will save $63.00 per year by using cold water according to Alliance to Save Energy in Washington DC and US Department of Energy and gives you the added bonus of longer-lasting clothes. Between 80 and 85 percent of the energy used to wash clothes comes from heating the water. Carbon savings are about 450+ pounds per year. Typically 400 loads of laundry are washed in US homes each year. Each full load in a standard washer uses 40 gallons of water.

Turn off water while brushing your teeth: This tip **$aves three gallons** of water each day per

person. On average 100 gallons of water per person, are used daily for hygienic needs. Carbon savings can easily reach 250 pounds per year. Environmental Defense Fund: http://www.fightglobalwarming.com/documents/5120_BrochureR4.pdf
US Environmental Protection Agency: http://www.epa.gov/safewater/kids/water_trivia_facts

Speaking of pearly whites, unplug that electric toothbrush. It is true that the electric brush does help prevent dental decay and gum disease but you don't need to have a hot handle 24/7 to do so. Most brushes will hold a charge for three days. Plug it in for about eight hours every three days.

Turn off water: Did I say that already? Now, I'm talking about when you shave. Just fill a shaving pan or the sink with a small amount of water to rinse your razor. This **saves three gallons** each day per person. Carbon savings are about 350 pounds per year and even more if it is hot water.
Environmental Defense Fund:
http://www.fightglobalwarming.com/documents/5120_BrochureR4.pdf

Take shorter showers: Turn off water while soaping, shaving, scrubbing then rinse. Heating water is one of the top energy expenses in the home.
Here is a tip to decrease this expense. Let's look at the numbers. If you take a five minute shower with a low flow 2.4 gallon per minute (GPM) showerhead instead of a ten minute shower with a 3.5 to 6.0 GPM (standard) showerhead, you'll save 23-48 gallons of water per person. That adds up quickly at 365 days X 23 gallons (2.4 GPM) = 8395 gallons; or 365 days X 48 gallons= 17,520 gallons. Now, multiply by the number of people in the house. If you do the math you will see the obvious. Get a five-minute shower timer if necessary and start taking shorter showers. Say, can you say, "soap, scrub, shave, shower and SAVE. No sh___"? You're definitely on your way to the bank.

Shorter Showers from Bates Motel

Turn off exhaust fans: When cooking or bathing in the winter, exhaust fans are a source of heat loss. They remove water vapors and humidity that is valuable to the home.

Disconnect: Unplug the electric garbage disposal to **save 50 to 150 gallons** of water per month, several kilowatts of energy and then start composting. Please, with or without a garbage disposal, do not put grease or oil down the drain. Grease does the same thing to sewers as cholesterol fat does to the human blood vessels: clogs them up, breaks the pipes, spills over and destroys everything in its path. This stuff is so bad some cities have ruled no garbage disposals can be placed in new construction. When this happens in the human body it results in a stroke or heart attack. This is just a snap shot in time of the same effects in the sewer system. Groovy Green: http://groovygreen.com/groove/?p=1232

Defrost frozen food: Defrost in the refrigerator before cooking. Don't be tempted to "just run it under warm water until its ready." That wastes energy to heat the water and wastes the water too. Obviously, this little tip requires thinking ahead...Hmm... things are getting more difficult now. Flip your frozen package over, turn it around but don't use hot water or the microwave to defrost.

Boiling and cooking: Bringing soups, sauces or water to a boiling point is more efficient on a stove top than in a microwave because it requires five percent more power to boil in the microwave. This is only boiling, though. It still requires less energy to warm a cup of coffee to perfect sipping temperature in the microwave. And generally, for quick warming or low heating, the microwave, toaster oven, pressure cooker or crock-pot is more efficient than stove top heating. Boiling, broiling and extreme high temperatures as in baking are more efficient in the conventional oven as is the range for is for a rolling boil.

But getting back to boiling-of-liquids topic: put a lid on it. If you would like to play scientist just try this experiment. Boil the identical amount of water in two identical size pans; one with a lid, one without. You will be amazed at the length of time it takes to go hmmm topless!
Some more tips about stovetop conservation:

- Place the pot on the burner that is closest to the pot's size.
- Cut vegetables to the smallest size because they cook faster.
- Another less known tip is to start out at the highest heat and then adjust downward when boiling.
- Remember, especially in warmer months, to keep the steam in the pot *(with the lid on)*...not in the kitchen.
- In winter, when you're finished with the boiling water, don't send it down the drain: let the steam humidify the air and it will add a degree or two the kitchen's ambient temperature. After it cools, this water can be used to water those houseplants.

The water is usually loaded with vitamins. Plants love it.

Did you know if you bake with ceramic or glass bakeware, you can decrease your oven temperature down by 25 degrees? This helps to save additional energy. And remember all the common things such as arranging the baking shelves before you turn on the oven, don't open the door to peak at the food, and turn the oven off a few minutes before the desired time especially in an electric oven or cook-top. It will still keep cooking for several minutes. Oh, and please don't bake in the months you are using your air conditioner. It is less expensive to purchase those brownies at the bakery than heating the oven and cooling the house...yes, even for chocolate!
Eco Kids:
http://www.ecokids.ca/pub/eco_info/topics/energy/energy_efficient/energy_quiz.swf

Close the fireplace damper: Keep your fireplace damper closed unless, of course, a fire is crackling. Keeping the damper open is similar to keeping a 48-inch wide window open during the winter. Just because it is out of site, don't let it slip out of your mind. All that expensive warm air will just float up and out of the chimney. Heat always rises to any occasion. An open damper will cost you approximately 14 percent more on

~GREEN BEINGS —SAVE GENES~   Page 43

your heating bill. This reminder comes from the US Department of Energy.
http://www1.eere.energy.gov/consumer/tips/pdfs/energy_savers.pdf

Night-night kilobyte: Set your computers to sleep and hibernate. Enable the "sleep mode" feature on your computer, allowing it to use less power during periods of inactivity.

- In Windows, the power management settings are found on your control panel under power options icon.

- Mac users, look for energy saving settings under system preferences in the apple menu. Configure your computer to "hibernate" automatically after 20 minutes or so of inactivity. The "hibernate mode" turns the computer off in a way that doesn't require you to reload everything when you switch it on.

- Allowing your computer system to sleep for 12 out of every 24 hours would save 300-576 pounds of GHGs annually. It is more time-efficient than shutting down and restarting your computer from scratch. This will reduce your household electric rate by up to two percent. Natural Resources Defense Council: http://www.nrdc.org/air/energy/genergy/easy.asp Alliance *to Save Energy Power$mart Booklet.*

Phantom of the appliances*:* Unplug electric devices: Phantom electricity costs the average US homeowners about 50 watts of electricity per month. Most appliances like TVs, VCRs, DVDs, printers, computers, cell phones, etc. are ready powered and prove it with their indicator light. They use from one to five watts per hour 24/7. They require stored energy to instantly come on or alive at your request and use electricity even when they are turned off.

Let's take another look at the numbers. Count all your electric sockets. How many are in use? Most folks can easily find 15-20 items. Fifteen devices at five watts per hour equals 75 watts, now multiply that by a twenty-four hour day. At my electric cost of 13.8 cents a kilowatt it would cost me $.25 +/- a few cents to keep those items plugged in. Let's see my annual cost would be almost $90. Wow! I just surprised myself.

By unplugging those electronic appliances, phantom energy loss will be halted in its current. According to Climate Crisis, five percent of all domestic energy use is

wasted this way. That translates to 18 million tons of carbon into the atmosphere and $8 billion annually.
Union of Concerned Scientist:
http://www.ucsusa.org/publications/greentips/energy-vampires.html
—*Lawrence Berkeley National Laboratory and quoted in Alliance's Power$mart booklet*

" HE IS RESPONSIBLE FOR THE PHANTOM ELECTRICITY USE."

Save 50 percent with the flick of a switch: This is an easy one…adjust your computer printer to print on both sides. Wow. You just doubled your paper use and saved more trees. Remember to purchase only recycled paper.

Empty nest? Save on heat and air conditioning on rooms not being used. If you have an extra room not currently being occupied, close the vents and close the door. Hang a sign on the doorknob to 'keep closed'.

Know when to open 'em ♪ know when to close 'em: You can achieve better heating and cooling from your home's windows if you know how to use sunlight wisely.

During the heating season in the cooler months, leave shades and blinds open on sunny days, but close them at night to reduce the amount of heat lost through windows. Close shades and blinds during the summer or when the house is heating up.

- According to the University of Florida, medium-colored draperies with white-plastic backings have been found to reduce heat gains by 33 percent.
- To reduce heat exchange or convection, draperies should be hung as close to windows as possible. Also let them fall onto a windowsill or floor.
- For maximum effectiveness, you should install a cornice at the top of a drapery or place the drapery against the ceiling. Then seal the drapery at both sides and overlap it in the center. You can use Velcro or magnetic tape to attach drapes to the wall at the sides and bottom. If you do these things, you will reduce additional heat loss.

More on windows: Open windows in the evenings in the spring and summer. Then close them in the morning to keep the cool air in. If you have windows that lower from the top too, it is better to open the window at the top and let the heat that naturally rises to the top of the room out of the house. Do this especially above the first floor, not while using the a/c though!

Oh, Don't get blue: Before jumping in the shower capture that cold water for your plants. You don't have to waste the water filling the pipes between your water heater and your shower. Instead, while waiting for the perfect temperature just place a bucket or jug under the faucet. You will save **200 to 300 gallons** of water per month. Probably even more when you consider you can use it on your inside or outside plants or to flush your toilet. Keep those empty juice gallon containers handy. Your plants will be glad and so will your pocketbook.

"Moon, be glad you are out of this atmosphere. They are killing me with this air. I continue to experience huge waves of sweats, then quaking chills, followed by dry hot fevers and persistent erupting vomiting. My entire equilibrium is so off balance. Please help me!

Get involved: If you are a US citizen you have a right to vote. Remind your congressman and delegates your vote could be the most important one in their next election. Contact your legislator by calling 1-800-491-2-7122. Here are some pointers when you call:

◆ Tell your legislator right off that you live in their district--that makes you a *constituent* (someone who can vote for them).

~GREEN BEINGS –SAVE GENES~

- Be focused and keep your message simple and short. Be *specific* and ask them to take an action such as voting for or against a bill.
- Keep it *brief* but be *personal* and always be *polite* and respectful, thanking them for their time.

And, even on the telephone, do this FREE tactic: SMILE. It has been proven that it takes less energy and a smile crosses telephone lines as well as in emails too. Smile!

Expand your horizons: There is an abundance of motivational media to choose from on the topic of energy conservation, climate change and solutions to both. They are full of facts and figures and helpful hints and resources too. Videos such as: *An Inconvenient Truth, Kilowatt Ours, Who Killed the Electric Car?* And *The Future of Food*, are a few titles that are interesting and inspirational. They show us what is really possible if we put our minds to it. Think about holding movie nights for your community and spread the word.

Chapter Three

Driving Cents to the Bank

Woof, who's got the gas? In the US currently nine million barrels of gasoline are consumed daily. That happens to be 44 percent of total global daily gasoline consumption. Price doesn't matter on this one. $1 or $4 per gallon, gasoline pollutes just the same. In fact, one gallon of gasoline releases 28 pounds of CO_2 into the atmosphere and increases ozone levels. Ozone traps heat and everything suffers! It is our pocketbook though that's catching up with the other climate and personal health changes that finally have captured our attention.

If the average miles per gallon on new cars were raised to a minimum of 40 mpg, there would be a savings of three million gallons of oil every single day. Fuel economy is so important because you can personally save between $200-$1500 per year by choosing efficient autos and altering your driving habits. No need for all those little paper receipts either when you fill up.

Fuel economy would decrease foreign imports of petroleum and crude oil on which the US spends over

$4.5 billion each week for imported foreign oil. Eventually it will be depleted or we will be deleted.

While the public waits for better automobile production, we all need to decrease our dependency as much as possible by making some of the changes discussed in this book. Check out this site for the lowest gas prices in your city and state: US Department of Energy: http://www.fueleconomy.gov/feg/gasprices/states/index.shtml

Stay Alive, Drive 55: Here we are again, back to Jimmy Carter's wise leadership. Driving slower saves $$$. Obey the speed limit. It's safer and less expensive. This is probably the most difficult habit to change, but you will be immediately rewarded. Gas mileage decreases rapidly above 60 mph. As a rule of thumb, each additional five miles per hour over 60 mph is like paying an additional 20 cents a gallon for gas.

If you like a challenge this can be fun, especially in traffic. Hold your speed at 55 mph or below according to limits and traffic. Watch car and truck drivers impatiently speed past, shake a fist, or wave with one finger. They will arrive at the next intersections before you. If you gauge your speed just right you will arrive, as the light turns green and can smoothly sail right past them, almost without braking. Watch them quickly accelerate again and see the dollars and frustration fume from their exhaust. This same pattern occurs over and over. It is amusing. Amusement decreases stress. Now, that is a safe and efficient way to get to work.

Consumer Reports found that cars increase their gas mileage five mpg or 15 percent when the speed is decreased by ten mph. In addition more than 1,500 pounds of GHGs will be spared from the environment by slowing down ten mpg. So, vehicles gain an additional five mpg when traveling speeds drop from 75 to 65 mph. The slower you travel the more miles per gallon you'll "earn". Alliance to Save Energy and the Environmental Defense offers these statistics on speed of travel and fuel consumption. Since fuel is 57 percent higher now than $2.30, you can almost double the following figures.
http://www.fightglobalwarming.com/page.cfm?

tagID=263 http://www.ase.org/content/news/detail/3713

Average gas mileage	Average fuel used (based on 12,000 miles per year)	Approximate greenhouse gas pollution	Approximate Cost (based on $2.30/gallon)
50 mpg	240 gallons	2.7 tons/year	$552
40 mpg	300 gallons	3.4 tons/year	$690
30 mpg	400 gallons	4.5 tons/year	$920
25 mpg	480 gallons	5.4 tons/year	$1,104
20 mpg	600 gallons	6.8 tons/year	$1,380
15 mpg	800 gallons	9 tons/year	$1,840
10 mpg	1,200 gallons	13.6 tons/year	$2,760

Here is how you determine your miles per gallon: The next time you fill your gas tank, reset your trip odometer to zero. Drive as usual for one tank-full of gas. At the following fill up, read the trip odometer miles recorded, for example 337 miles were driven and one-half tank gas was used. Look on the stations gas pump just under your costs, to read the number of gallons used. On a recent fill up 10.5 gallons were used to drive 337 miles. Simply divide 337 by 10.5 and you get the most accurate miles per gallon for your vehicle. In this case, the five-year-old Saab got 32 miles from each gallon of gas just for following the simple tips listed in this chapter.

Let Tom do the driving. Cruise that is like a Sunday afternoon: Using cruise control helps to cut fuel consumption as well as controls the desire to put the pedal to the metal. Cruise control maintains a steady speed during highway driving. Driving slower saves money. You will be rewarded with twenty percent more in miles per gallon of fuel. Depending on your vehicle, most autos will get an extra 100 miles free on every tank of gas just for driving slower. Truly, you don't need a special hybrid car or particular kind of fuel to improve your immediate gas mileage…just plain old cruise control with a dash of self-control.

Plain and Simple Common Sense: Paying $80 or more per tank of gas probably allows you to drive about 400 miles before reaching empty. Driving slower stretches your mileage to 500 or more, for those same costs. You decide if the extra five minutes you gain from speeding is worth the expense. Example: if your car normally gets 20 mpg and you change your driving habits, you will gain an extra 100 miles. That is like getting 5 gallons of free fuel. And at $4 gallon that equals $20 saved per fill up. If your auto has cruise control, try using it for one tank-full. I promise you will be thankful. Set your miles per hour (mph) at 55.

To save even more, keep your engine's revolutions per minute (RPM) less than two (thousand). You will need to go on-and-off cruise control to maintain the RPMs. If automakers were smart they would design cars with RPM, not MPH cruise control. The RPM saves more fuel because the car speed will slow going up hills to keep from using more fuel. The engine and not the speed, stays steady and smooth on hills and plains. Alliance to Save Energy: http://www.ase.org/content/news/detail/3713 As reported in the *Wall Street Journal* fuel-saving gadgets priced from $35 to $300 contributed to 4 percent auto aftermarket sales. None are proven to be as effective as the simple tactic listed in this chapter and endorsed by the EPA.

Keep on rolling: By keeping tires fully inflated, you save at least $100 per year. For proper tire inflation - check your owner manual or sticker on the inside of the front driver doorframe for the proper amount. Keep a tire gauge in your auto's glove box.

Under-inflated tires will increase rolling resistance and decrease mileage. When your tires run too low on air, the sidewall bows out and the tread flattens on contact. While this adds grip on sand, fine gravel, mud, and snow, it causes your car to "wallow" more on turns, makes it easier to hydroplane, and in extreme cases can actually cause a tread separation! The flexing of the rubber and threads heats up the tire and provides more rolling resistance, which is bad for tire life and fuel economy. Proper inflation of your tires also improves gas mileage by more than three percent when maintained regularly. Keep this in mind: under-inflated tires can lower gas mileage by 0.4 percent for every one pounds per square inch (PSI) drop in pressure of all four tires. In addition, did you know that tires could lose about one PSI of pressure for every 10 degrees of temperature drop? So you must check the pressure more frequently in the cooler months.

Keeping your car properly tuned and oiled will save 580 pounds of carbon each year while gaining another six percent improvement on your fuel use. Check your tire pressure every 4 weeks and inflate them to the recommended level. Car Care Aware Canada: http://www.carcare.org/Tires_Wheels/inflation.shtml Alliance to Save Energy: http://www.ase.org/content/news/detail/3713

Yea, right! Here is a tip from United Parcel Service, stay out of the left turn lane while driving in heavy traffic. It is true making three right turns versus one long left turn lane can save lots of fuel. In fact it has saved UPS 3 million gallons of fuel in 2007 and reduced carbon emissions by 32,000 metric tons. We can learn from them and keep on saving money. Right on!

"When are they going to learn burning fossil fuel just isn't going to allow their transportation system to ever progress or catch up with us....stupid humans"

Plan your route: This is amazingly effective. With just a little bit of thought you can accomplish those necessary errands on the way to or from work. Consider using a global positioning system (GPS) devise for precise locations when traveling to a new location. With the climbing price of gas, the GPS will pay for itself quickly. Another option is a three-dollar per request or the ten-dollar per month navigation service fee that some telephone providers offer. Not only do you save money, you prevent that nagging stress when lost or late. One won't need to stop for directions ever again!

A free alternative to this is **Map-Quest's route planner.** This allows you to plan ten stops, arrange and rearrange them until you are satisfied with the shortest or quickest route. Map-quest can also show you the prices of gasoline at different stations and locations before you arrive. It is EEEEZZZZY! http://help.mapquest.com/jive/entry!default.jspa?categoryID=4&externalID=344&fromSearchPage=true or http://www.mapquest.com/directions/main.adp?bCTsettings=1

Carrying too much baggage? Are you carrying around too much excess "baggage?" Improve your relationships and your gas mileage too; empty your trunk, truck, back seat, roof rack, trailer hitch and all extra weight. Pack lightly when traveling and avoid carrying additional weight. Every additional 100 pounds of excess weight in your vehicle costs about two percent more in fuel consumption. Alliance to Save Energy: http://www.ase.org/content/ news/detail/3713.

***Stay* out of the devils workshop;** Devil will stay out of your details. Don't idle. If you are stopping you car for more than ten seconds—except in traffic—turn off your engine. Idling for more than ten seconds uses more gas (i.e., more $$$) and creates more global warming pollution (GHG emissions) than simply restarting your engine. This means not starting your engine until you are ready to put it in gear and drive away. Put your safety belt on first, arrange your mirrors, your seat, set your radio or arrange your person items before starting the car. It also means turning off your engine at bank or fast food restaurant drive-up-windows, long lines at toll-bridges or the quick run into the cleaners and especially while waiting to pick up someone. Eliminating five minutes of engine idle time per day for one month will reduce your CO_2 emissions by 27 pounds. At $4 a gallon, and if idling the car for one hour uses one gallon of gas, each minute you idle will cost you seven cents. Hang a 'no idling' sign as a reminder on your auto's mirror.

Brown (reusable) bag it: If you have been driving out for lunch from work try taking your lunch. It is healthier and less expensive, especially when you add in the additional cost for gasoline, already at $4 per gallon. P.S. use a canvas or reusable bag even if it isn't brown.

Be slow cool: According to American Automobile Association, the magic number is 40...miles per gallon, that is. If you are driving over 40 miles per hour, use your car's air conditioning in the summer. If you are driving under that, you'll use less gas if you open the windows to cool down. The reason is the extra drag you put on the car by opening the windows at highway speeds uses more gas than the air conditioner. Save even more by tinting your car windows, installing portable (solar or rechargeable battery) fans for the auto dashboard and park in the shade whenever possible. All this from Jelsoft Enterprises LTD: Saving Advice Dot Com: http://www.savingadvice.com/forums/money-saving-tips/61-gas-saving-money-tips.html.

This one's for the boys: Clean your car's air filter. Standard maintenance for most cars recommends replacing the air filter every 15,000 miles. Did you know the car's air filter begins to get clogged with dust after just a few thousand miles? As the filter becomes more clogged, the airflow under the hood slows down and your automobile becomes less fuel-efficient. How much less efficient could that be? A dirty air filter, even after just 5,000 miles can represent about seven percent of your fuel cost. If you are driving a car that normally gets 22 miles per gallon, your car is now getting 20.5 mpg. Over the next 10,000 miles, that's an extra 33 gallons of gas and at $4 per gallon it will save you $112.80! Solution: pull up the hood and vacuum your filter each time you wash your car. The Simple Dollar, http://www.thesimpledollar.com/2007/02/19/five-minute-finances-1-clean-your-cars-air-filter.

~GREEN BEINGS —SAVE GENES~

Online banking: Online banking save stamps, envelops, paper, trips to the post office and gas for the US Postal Service because they don't have to carry the additional weight of your bills. Get paperless bank statements too. If every home in the US viewed and paid bills electronically, the country would save 18.5 million trees and avoid two billion tons of toxic air pollutants. Here is another very scary thought...many Banks of America offer *services in Braille at their drive-up windows!* Go figure! Don't drive...bank online!
http://www.nyc.gov/html/planyc2030/html/greenyc/greenyc.shtml
http://www.bankofamerica.com/onlinebanking/

Debunked: Conventional wisdom of the 3,000-mile oil change is another myth. Most cars produced in the past five years recommend an oil change between five and ten thousand miles. For example, the 2003 Honda Civic recommends 10,000 between oil changes. Vehicles ten years old or more may require more frequent changes. But newer vehicles don't need to change their oil as often even though many businesses go on recommending doing so. http://www.3000milemyth.org. Here are some important tips:

- ◆ Check your manual for the frequency of an oil change for your car

- ◆ If you don't have the manual, check Google. Within seconds you will have the answer

- ◆ Do not change the oil too often. Save money here too.

- ◆ If you change the oil yourself, recycle the oil and the filter. www.recycleoil.org

- ◆ Keep kitty litter on hand to soak up any spills

- ◆ Most importantly keep oil and other fluids away from storm drains.

- ◆ Never dump anything, especially oil directly into local waterways, creeks, rivers, bays or the ocean. Oil and water equal disaster. links.sfgate.com/ZCJP.

" I TOLD YOU NOT TO DUMP ALL THAT OLD ANTIFREEZE IN THE OCEAN"

Easy for Einstein to understand: Fill your car in the morning: Sound silly? Here is an interesting fact to help save even more money. In the morning the ground temperature is actually cooler. This affects the temperature of the gasoline in the underground tank. When it is cooler the result is a denser product, not dumber...denser. In the US gasoline pumps are measured by volume. This actually means you'll get slightly more gas per gallon in the morning then the evening when the ground tanks and gasoline have had a full day of warmer temperature. A one-degree rise in temperature is a big deal. You will be able to appreciate a slight increase in mpg.

Here's a good visual to understand this concept: Think about an ice cube container. Fill it to the brim. When it freezes it shrinks and becomes smaller and denser. You could actually add a bit more water to the same container. It is just the same with fuel. When it is cooler you actually get more for your money. The Simple Dollar: http://www.thesimpledollar.com/2006/12/07/ten-ways-to-save-money-on-your-evening-commute

Don't squeeze the Charmin - or the gas nozzle - too hard: Really, if you take the time to notice, gas nozzles have three speeds of operation. Most of us squeeze hard to get gassed up quickly. This is not economical and here is why: in the slowest mode the gasoline flows low and steady without emitting vapors. If you squeeze hard and fast, the gasoline leaves the hose with turbulence and result in vapors or fumes coming from the liquid. And even more infuriating is that some of those vapors are being sucked right back into the underground storage tank where they return to liquid. So, you are, in effect, paying for fuel that you just gave back to the gas station! Just like driving, go slow for more efficiency.

Depending on your circumstances, these tips may be helpful. If you have purchased a $35 membership to

Sam's Club you will receive an addition 3 percent discount on their gasoline. If you have a basic $62.50 membership to Triple AAA auto club, you may apply for their fuel credit card to receive an additional 3 percent discount on fuel. With both discounts the average driver will save about $100 per year. Since this is the cost of the membership it is not worth purchasing for fuel discounts alone, but there are plenty of other benefits too.

Another mystery explained: Have you ever noticed that the first half of the gas tank seems to last much longer than the second half? In fact, the second half almost seems to disappear in a flash. If we go back to the prior vapor story, it will be easy to understand why this is. Gasoline evaporates when it is exposed to air. As your tank becomes empty there is more air contact available in the tank and less fuel. The moral of this story: to put more pennies in your pocket fill up your gas tank before it reaches the halfway mark. *Ode* magazine, http://www.odemagazine.com/doc/48/making-the-golden-years-shine

Jump, dive or just flip in the pool: Join a carpool, ride a commuter bus or ride a bike to work. Try it just one day per week to make sure you won't drown. It costs about .31 cents (and rising) per mile in an

automobile for fuel and maintenance. Not only will commuting together save you money, reduce congestion and cut carbon emissions, you will actually meet new friends, too. Prove it to yourself...money talks. That is yet another reason to forget Geico when you carpool. This habit will save you thousands of dollars. Shoot, it is like getting gasoline for $2 per gallon by just adding one rider, and hmmm $1.33 per gallon of gas (based on four dollars per gallon) if you ride to work with two people instead of one. Check out Ride Share online and this site will provide your personal savings when carpooling. Calculate your potential savings.
http://www.greenoptions.com/blog/2007/03/13/tip_o_the_day_i_hov_you
Environmental Defense:
http://www.fightglobalwarming.com/page.cfm?tagID=268

http://www.sfgate.com/cgi-bin/article.cgi?f=/c/a/2008/02/10/BA1EUUCL9.DTL

Recap:

- Cruise at 55 mph or keep RPM at 2x1000
- Don't idle
- Always keep fuel tank ½ full
- Fill fuel tank slowly and in the morning
- Keep tires fully inflated
- Plan your errands around your commute to and from work
- Change oil every 10,000 miles
- Vacuum your air filter monthly
- Stay out of the left turn lane
- Remove excess weight from vehicle
- Ride with a friend or two
- Reap the benefits; less fuel costs, less vehicle insurance expenses, less auto maintenance and more friends, and perhaps a longer life.

Chapter Four

Dialing for Dollars

Slick idea: Good job...you saved a few bucks changing your oil at home...now what do you do with the oil? Call 202-682-8000 or visit http://www.recycleoil.org or http://earth911.org/ (Perhaps you will want to do that before the oil change). Did you know refining used oil takes 50-85 percent less energy than refining crude oil? Keep up the good work!

If you don't ask...they won't tell: Check your utility company for incentives. They often offer special rates to encourage you to use power at off-peak times. By using appliances at low-use hours you may save 20-30 percent off the price of a kilowatt for that device. Instead of paying 13.8 cents per kilowatt-hour, you may only be charged eight cents per kilowatt just for starting your washer and dryer late in the evening. It all translates to money in your pocket and carbon spared from the air. Timing is the key here. Even without any lower costs, by using electricity at off peak hours you are actually helping to prevent the construction of more power coal burning plants. You see the decision to build is based on peak demand needs. Avoid using electricity at peak times a much as possible.

~GREEN BEINGS —SAVE GENES~

While you are calling ...check for a free programmable thermostat: Again some utility companies offer customers a choice of either a free programmable thermostat or an air conditioner load control switch that will help reduce energy consumption and costs. This technology provided by participating utility companies allows them to cycle air conditioning heat pump use during periods of very high electricity demand. Many cities have similar programs. Baltimore Gas and Electric Company's adopted this program in 2007. Check your power company for a similar program.

Prevent clutter, save a tree from a bad day: Cancel all junk mail and catalogs. Every year Americans collectively receive four million tons of junk mail. This mail comes directly from 100 million trees. Save our resources. Just about 50 pounds of paper will save one entire tree. Much of the tree pulp is wasted in processing. Visit Direct Marketing Association's http://www.the-dma.org/consumers/index.html. You must complete the form, pay $1 and your name will be removed from most mailing lists. Or you can call 1-888-5-opt-out to remove your name from repeated junk mail offers.

For store catalogs, contact this free service to stop them at your home and or office. They make it so easy. In addition every catalog is available online for your shopping needs. Remember, when you shop online you are decreasing your gasoline expenses by not driving mall to mall.

Finally, you should rid yourself of all those frequent and bulky local fliers. These are the fliers that are frequently sent to *Current Resident*. To stop delivery please contact Green Options Dot Com or ADVO:
http://www.catalogchoice.org/#welcome
http://www.advo.com/aboutadvo.html
http://www.greenoptions.com/blog/2007/01/14/tip_o_the_day_please_mister_postman_no_more_junk_mail

Free call stops credit cards and mortgage offers for five years: Call 888-567-8688, an automated phone line maintained by a major credit bureaus. Stop the 500 pieces of related junk mail you receive each year. If any remaining junk mail you can contact Direct Marketing Association Mail Preference Service at P.O. Box 643, Carmel, NY 10512

The real yellow pages: This is a huge waste of resources. Please cancel home delivery of all the telephone books delivered to your residence. The information is available online or on cell phones and is an unnecessary waste of precious resources. Discontinuation of service is on the first few pages of the books. Often the number to call is the same easy-to-find number listed for placing an advertisement. Try this number in the Northeast 1-800-373-3280

Have green will travel: This one is from Robert and for all those traveling folks out there. Opt for green hotels when you can and let lodging management know you don't need clean sheets and towels every day unless, of course you do that in your own home. Most rooms provide an index card in the room to select this option. Or call the Green Hotel Association.www.greenhotels.com Phone: (713) 789-8889, Fax: (713) 789-9786, P.O. Box 420212, Houston, TX 77242-0212

A responsible hotel chain to support is Marriott. They have recently committed to protecting one and a half million acres of rainforest and $2 million in energy conservation in their 3,000 hotels. Though they admit their intentions are not solely altruistic (they do it to save money) they realize that a two-foot rise in sea level will destroy many of their coastal properties.

Eco-Maps would love to tell you where to go: This innovative mapping project started in 1995 has more than 350 maps available online representing more than 50 countries. Be sure to check out local green and farmers markets, thrift stores, greenway projects, bike paths and places to compost your trash while traveling. No reason to give up your good habits while traveling when green maps are available. http://www.greenmap.org/

Free fish information: This amazing database from Blue Ocean Institute allows the consumer to make intelligent decisions on choosing the healthiest fish dinner. Besides loss of species from over fishing, some fish contain harmful levels of mercury, PCBs, dioxins, or pesticides. The information is easy to understand with their descriptions. Visit their web page prior to grocery shopping or restaurant dining. Or, for even more convenience, the same information can be obtained from your cell phone with text message capabilities. Simply call the toll free number below and enter the name of the fish you wish to consume. In seconds, you will receive information from the following rating system.

🐟 Species is relatively abundant, and fishing/farming methods cause little damage to habitat and other wildlife.

🐟 Some problems exist with this species' status or catch/farming methods, or information is insufficient for evaluating.

🐟 Species has a combination of problems such as over fishing, high by catch, and poor management, or farming methods have serious environmental impacts.

✅ A fishery targeting this species has been certified as sustainable and well managed by the Marine Stewardship Council. Learn more at www.msc.org.

🚩 These fish contain levels of mercury, PCBs, dioxins, or pesticides that may pose a health risk to adults and children. Please refer to www.environmentaldefense.org/seafood for more details.

🐟 Here is a sample report on one of my favorite fish. "Orange Roughy is severely depleted. Orange Roughy don't mature until they're at least 20 years old. If not caught, this fish can live over 100 years. They live in deep waters where habitat-damaging trawlers catch them when they gather in groups to feed or spawn. Fishing for Orange Roughy also catches and kills a number of threatened deep-sea shark species".

This is one way to help restore abundance in our oceans. Toll-free 877-BOI-SEAS or E-mail: seafood@blueocean.org or visit www.blueocean.org

Chicken of the Sea

Chapter Five

Garden Delight

Free garden fertilizer: From Starbucks for your garden, coffee grounds and they have plenty of this stuff. Coffee grounds contain an abundance of nutrients like nitrogen (1.45 percent), potassium (1240 micrograms), calcium (389 micrograms), magnesium (448 micrograms) and it is high in sulfur too. http://www.starbucks.com/aboutus/compost.asp And for 21 other ways to use old coffee grounds from repelling ants to skin and hair treatments, visit http://www.greendaily.com/2007/12/28/21-ways-to-use-old-coffee-grounds

Non-toxic lawn fertilizers: At some point in your gardening experience, you will need a fertilizer to give your plants an extra boost. Unfortunately, many gardeners rely on chemical fertilizers to grow their trees, grass, flowers, fruits, and vegetables.

Thankfully, there are a number of healthy natural and organic fertilizers available that will help you grow a lush, green garden and lawn without the use of potentially harmful chemical fertilizers. Both the traditional and organic fertilizers show their content with three bold numbers. These numbers represent three different compounds and percentages: Nitrogen (N), Phosphorous (P), and Potassium Potash (K). For example (N) helps plant foliage to grow strong. Roots and flowers develop better with (P). Overall plant health is encouraged (K).

Organic fertilizers are made from a huge variety of naturally occurring elements besides coffee grounds. Things such as bat guano, bone meal, feather meal and fishmeal are all loaded with NPK. Check out the USDA National Organic Program for more information about organic gardening products and standards http://www.cleanairgardening.com/fertilizeguide.html or try http://www.gardensalive.com/Default.asp?bhcd2=1209733825

Grow (don't smoke) great grass: Here is a favorite resource for 2,168 natural solutions when growing just about anything. Not only tricks for encouraging plant growth but solutions for discouraging all the critters that can attack our lawn and gardens are offered by using everyday ingredients like, beer, tobacco, Tabasco sauce, vinegar, moth balls, tea, even Karo syrup has a place in the long recipe list. Read Jerry

~GREEN BEINGS —SAVE GENES~

Baker's *Backyard Problem Solver*. He is the guru of home remedies for old-fashion gardening.

Just the facts, please: Take a guess on this one: who pollutes more a vegan or a driver of the Prius hybrid automobile? The answer is the hybrid driver is responsible for one ton of green house gases per year. The vegan (one who eats no dairy or meat products) is responsible for zero green house gasses arising from animal sources. Meat eaters are responsible for 1.5 tons of green house gas per year. Bottom line is eating less meat is the smart thing to do.

Now that you have the idea, think about this one. Who is responsible for less carbon emissions-the energy efficient suburban mini-mansion home owner with a Prius automobile, or a city dweller in an old less efficient abode living only miles from the place of employment? P.S. it's the city slicker.

Free insect disposal/eater: Build a bat box and hang it high where it gets morning sun. Bats eat between 600-1000 mosquito-sized insects every hour. Get free plans to build the bat box at: Bat Conservation International Inc: www.batcon.org and or Organization for Bat Conservation: www.batconservation.org

~GREEN BEINGS —SAVE GENES~

Collect rainwater in barrels: Lawn and garden watering make up nearly 40 percent of total household water use during the summer. Just one rain barrel will save most homeowners about 1,300 gallons of water during the peak summer months. It would be nice to have one on each corner of the house. Saving water not only helps protect the environment by preventing run-off. It saves you money and energy (decreased demand for treated tap water). Maryland Environmental Design Program: http://www.dnr.state.md.us/ed/rainbarrel.html

INDIVIDUAL ACTION

Medusa and worms love this stuff...composting: This is the safest way to return our daily waste back to Earth. A compost bin helps to speed up the break down or decomposition of all natural waste. When organic materials are placed in a container with a tight fitting lid, microorganisms that tolerate high heat conditions begin chomping away at your garbage. Within weeks they create a rich humus type of soil that can be added to your lawn and garden. Iif you have the room, you can place a wire fence in a three-foot circle to contain the

~GREEN BEINGS —SAVE GENES~

collection placed directly on the ground. Don't forget to add grass clippings, leaves and other organic yard waste.

Keep a compost crock in your kitchen for convenience. All non-dairy, non-meat products can safely be composted. Do compost eggshells. The calcium in the shells is beneficial to the plants and will not attract wild animals. Composting just got easier with Bio-bags! These bags line your compost crock and make it easier to empty and clean. The bags feel like plastic but in fact are made from corn and contain no chemicals or additives. They will decompose along with the food waste.

No worries if you happen to live in an apartment or a town-home where no possible out-door compost bin is feasible. You can have a composter right in your kitchen. A nature mill can do the same thing without odors or insects. With the help of a few kW's, heat is added to start the process. One advantage to composting inside is that you can now add meat and dairy scraps without the worry of attracting varmints. Every few weeks just open the bottom compartment and find the food waste has been transformed to organic potting soil to use or sell. http://www.composters.com/compost-bins.php?gclid=CP754MHw2pMCFRcaagodYwdOZQ
Composting is good for the environment. Compost and CO_2 go together like salt and pepper. Not only does it have agricultural benefits, but composting also combats climate change. When plant wastes are sent to landfills they turn into carbon dioxide and methane, two of the

most common greenhouse gasses. When those plants are composted, they lock up carbon from the atmosphere for decades! And when you add that compost to your garden's soil, you are also sequestering additional carbon dioxide.

Green thumbs get greener in greenhouses with this gooey idea: This is a 15-second per window smart tip to cut your heating bill in your garden shed or greenhouse using bubble wrap, not bubble gum. But, you may want to use some bubble gum to attach your bubble wrap. After cutting the wrap to the size of the window, just to spray a film of water on the window and then apply the wrap to the wet window. It allows plenty of light in but will appear as if you are peering through a shower door. Reuse your packing material or get bubble wrap from retailers or companies receiving wrapped packages. Bubble wrap can increase the R-value 50 percent from 1 to 2 on single windows. Build It Solar: http://www.builditsolar.com/NewsLetter/newsletterArch0509.htm

Get fit and trim for free: The next time your yard needs mowing, just pull out the old hand-propelled push lawn mowerThey are quiet and require no gasoline. The new reel push mowers are lightweight, easy to handle, and produce no pollutants or greenhouses gas emissions and are modernized with scissor sharp blades and a grass collection basket. Besides the exercise one hour spent mowing a lawn with a gas-powered mower produces as many emissions as 50 hours spent driving an average car. Surprised? Now that is an air polluting dirty task! Not only will it save you on Kleenex ('aaaachooo!'), and fuel fees but also you won't need to pay that personal gym membership either. So, you can look good, feel better, save money and reduce pollution...which part didn't you understand? Wait a minute...what is the title of this book?

One more exercise tip...if you tire of your home exercise videos...you can trade it in exchange for a different one. Free! Visit www.videofitness.com. Here is another upcoming alternative: Biodiesel powered lawn mower from HUGR Systems Orlando Florida 321-299-8948. Or consider an electric mower especially if you have converted your home to solar or wind power for electricity.

Free home air filters: Let's talk about *power plants*. Want to improve the indoor air quality in your home or office? Get a houseplant. Plants can absorb air pollutants and alleviate some "sick building syndrome" symptoms such as headaches and eye, nose or throat irritation which may be caused by inadequate ventilation, chemical contaminants (VOCs, carbon monoxide) or biological contaminants such as mold or pollen. You don't have to overdo it. One plant for every ten square yards of floor space should be plenty. It is claimed that these houseplants are some of the best absorbers: Boston Fern, English Ivy, Rubber plant, and Peace Lily. Hugg: http://www.hugg.com/?page=2

Mary Mary, Harry or Larry how does your garden grow? Grow your own, plant herbs and vegetables in a garden plot or even in flowerpots. Homegrown foods are less expensive and healthier. Without using chemicals your food will be organic. Your plants will thrive by using natural composted soil or compost "tea" to fertilize. The tea is easy to make by just letting some of the compost seep in a large bucket of rainwater for about a week. The organic tea can then be diluted and sprayed or watered at the root to give your plants an extra heap of nutrients.

Not into gardening? That is ok you can still *buy organic*. Organic farmers don't use toxic pesticides or synthetic fertilizer. There are no genetically or chemically altered foods produced from plants and animals when certified organic. By going organic you will be impacting the health of your family, the planet and the farm animals. Keep your purchases limited to **certified organic food of the season.**

Equally smart is to 'buy local' foods. The average meal travels more than 1500 miles from farm to table. Visit the Organic Consumers Organization: http://www.organicconsumers.org they make it so easy for you to find local produce within 20 miles of your zip code. By doing so you are helping to decrease fuels costs in transporting groceries. You can personally make a large impact here. Another great web resource offering

organically grown food can be found here.
http://www.localharvest.org/

From turf to surf: The Center for Science and Public Interest has prepared a green food calculator just for you. http://www.cspinet.org/EatingGreen/calculator.html. Now you can see the actual effects of your food choices on your health and home. It is easy and it's free knowledge. Knowledge is power. Power is money. Stay tuned for more facts about carnivorous habits.

No whining about these grapes: If you can't stomp your own grapes, try this wine introduced to the US market in 2007. Fair Trade and organic wine has been produced in South Africa since 2003 and in Chile and Argentina since 2004. The South African certification process requires vineyard workers to maintain a legally protected 25 percent interest in the winery. The South African government's policies support and promote equal land ownership following the end of Apartheid. Fair Trade products are also free of harmful pesticides.

~GREEN BEINGS —SAVE GENES~

http://www.coopamerica.org/pubs/greenpages/results.cfm?category=&state=&keywords=wine

ORGANIZED CRIME

Strawberry fields of the 21st century

Get the free Fair-Trade guide: Here you can find coffee, tea, banana, rice, vanilla beans, spices, sugar, chocolate, wine, olive oil and many non-edible products too. If a farmer wanted to convert his crop to 'fair trade' certification, he would have to adhere to a long list of rules on pesticides, farming techniques, recycling and he must show that his children are enrolled in school. He receives a 20 percent premium for his efforts and so is able to better provide for his family. Read more on this topic in Chapter Eleven.
http://www.coopamerica.org/programs/fairtrade/orderguide.cfm

Growing US organic food sales to 10 percent by 2010 can improve our personal health by:

- Eliminating pesticides from 98 million daily US servings of drinking water
- Assuring 20 million daily servings of milk that are produced without antibiotics and genetically modified growth hormones
- Assuring 53 million daily servings of pesticide-free fruits and vegetables (Enough for 10 million kids to have five daily servings)
- Assuring that 915 million animals are treated more humanely
- Fighting climate change by capturing an additional 6.5 billion pounds of carbon in the soil (that's the equivalent of taking 2 million cars, each averaging 12,000 miles per year off the road.).
- Eliminating 2.9 billion barrels of imported oil annually (Equal to 406,000 Olympic eight-lane competition pools)
- Restoring 25,800 square miles of degraded soils to rich, highly productive cropland (An amount of land equal to the size of WV
- Lowering the number of pre-term deliveries each year (Prematurity is now an epidemic in the US, affecting one in eight babies. It is a leading cause of developmental problems, and death, in babies)
- Reducing unwanted interference with our sex hormones (This could reduce the prevalence of erectile dysfunction and the number of people

~GREEN BEINGS –SAVE GENES~

suffering from loss of sexual drive as well as other estrogen related problems)
- Lowering the incidence of neuron-developmental problems in children perhaps including ADHD and autism (Abnormal neuron-development in children can be caused or made worse by pre-natal and early life exposure to pesticides and chemicals that contaminate our food)
- Virtually eliminating dietary exposure to insecticides known to be developmental neurotoxins (based on the compelling findings reported in two University of Washington studies involving school-aged children)

http://www.coopamerica.org/programs/fairtrade/whattokno/12waystoshopfairtrade.cfm

"For in the true nature of things, if we rightly consider, every green tree is far more glorious than if it were made of gold and silver".
-- Martin Luther

Plant a tree: Poor Brooklyn had only "a tree": you can plant several trees. Here are some of the 20 excellent reasons from the Tree People to do so:

- ◆ Trees combat the "greenhouse effect" by Absorbing about 25 pounds annually of the CO_2 we exhale.
- ◆ As plants and trees grow, they absorb carbon from the atmosphere through photosynthesis and store it in wood, leaves, roots and soils.
- ◆ Most important, they release oxygen back into the air.
- ◆ This improves the air quality and prevents soil erosion by protecting the land and streams from runoff.
- ◆ Planting shade trees decreases cooling costs up to 50 percent and decreases up to 2.4 tons of CO_2 emissions per year.
- ◆ Trees in a city can decrease the ambient temperature by ten degrees.
- ◆ They protect us all from cancer causing ultra-violet rays. (Skin cancer is the number one increasing incidence due to loss of trees).
- ◆ Trees do so much more for us-providing food and wood, a gathering for friends and neighbors, and they provide a home to thousands of species.

- They make excellent fences and decrease noise pollution.
- They always increase the property value of a home or business.

http://www.treepeople.com/vfp.dll?OakTree~getPage~&PNPK=59

The Casey Trees website will allow you to compare many factors about the value of a variety of trees. Not only the carbon benefit is calculated, but the increased home value can be estimated too. What a great resource. http://www.itreetools.org/treecalculator/

Join The National Arbor Day Foundation and receive ten free trees guaranteed to grow. Remember plants and trees breath CO_2 that we exhale. So that is why they thrive on our talking to them. Here is another way to help offset carbon in the atmosphere. Donations as small as $10 will be accepted; for every dollar donated one tree will be planted. http://www.americanforests.org/ .

Chapter Six

Wee bit of Will Power

Solar dryer: Aka a "clothesline" is the best method to air-dry laundry. For those of you who don't have space for a clothesline in the backyard there is a remedy. You can set up a *stand-alone clothes rack* indoors. In the winter, you get a bonus of humidifying the air in your home without paying to operate another appliance such as a humidifier. Just air-drying alone will save you $80 a year and prevent over 1200 pounds of CO_2 going into the atmosphere. The humidifier uses 60 kW so additional savings will be appreciated.

When hanging clothing outdoors, Mother Nature outperforms bleach with her gift of sunshine. Spots and stains seem to magically lighten up or disappear. Laundry will smell fresh. Towels will feel even more absorbent than smoothies coming out of the dryer. Wait a minute: I think there may be more one benefit of line drying clothing.

There is growing legislation to overturn any previous bans on clotheslines. In fact new laws are pending in Connecticut, Vermont and Colorado-another reason to contact your local congressman.

~GREEN BEINGS —SAVE GENES~

Clothes dryers: If you must use a clothes dryer, make sure you clean the lint filter in the dryer after each use. Clogged filters decrease energy use by 30 percent. Dryers use 1080 kW or about a buck and a half for every load. So, if you can't line dry be sure to dry heavy and light fabrics separately. You know... heavy blue jeans, cotton socks and towels separate from that light pretty little camisole, permanent press sheet, shirts, etc. This is a no-brainer, don't add wet items to a load that's already partly dry.

If available, use the moisture sensor setting. A dryer isn't too much different from a refrigerator in its amount of energy us. As reported in the *New York Times* in their "line-in-the-yard" discussion, even though a dryer is used intermittently and a frig runs 24/7, the dryer still consumes more than six percent of your energy bills. Natural Resources Defense Fund: http://www.nrdc.org/air/energy/genergy/easy.asp

Be free...phosphate free: Avoid laundry or kitchen cleaners that contain phosphates. Phosphates appear in a number of cleaners such as many laundry soaps and nearly all dishwasher liquids or dishwashing granules.

- ◆ Phosphates rob the water of oxygen that fish require to survive.

- ◆ Phosphates promote rapid algae growth that pollute the water supply.

- ◆ They smell bad.

They are not needed and really were only added at the end of a war when phosphate supply was plentiful and no longer needed in weapon production. Now, we kill the rivers and streams with the same stuff because someone claimed it could boost the power of laundry soap but so does borax. In the last 60 years many streams and rivers have become dead zones because of chemicals like phosphates and because of agricultural runoff. They are scheduled to be banned by 2010 in dishwashing soaps. Why wait? Avoid phosphates now.

Companies such as Seventh Generation have been making non-phosphate laundry and dishwasher detergent since 1997. It is sold at many environmentally minded stores such as Trader Joe's and Whole Foods. Consumer report has investigated the phosphate free products and determined they do an excellent job.
http://www.forbes.com/2007/06/07/phosphate-detergent-ban-oped-cx_ae_0608ebeling.html
http://www.greenworkscleaners.com/products/definition.php

Make your toilet a low-flow bowl instantly: Toilets use the most water in private homes. They require more than 30 percent of all water use. To decrease the amount of water use in a toilet, install a plastic bottle filled with water or a plastic bag weighted with pebbles in your toilet tank. Or just put a brick in it. (That is in the top holding tan, please).

Displacing water this way allows you to use less water with each flush, conserving **five to ten gallons** a day. That's **up to 300 gallons** a month and even more for large families!

Check toilets for leaks. Put dye tablets or food coloring into the holding tank. If color appears in the bowl without flushing, there's a leak that should be repaired. There is another potential opportunity to **save 400 gallons** a month.

Finally if you can stand still long enough to fix those leaky faucets: Even a slow drip wastes ten to 25 gallons of water. Just think, 15 drips per minute add up to almost three gallons of water wasted per day, 65 gallons wasted per month, and 788 gallons wasted per year!

Toss out the bleach: When purchasing paper products-even recycled products, choose the natural color. Products that are bleached white contain dioxins. This harmful ingredient is well known to be carcinogenic as it accumulates in air, water and soil. Visit the Chlorine Free Products Association at http://www.chlorinefreeproducts.org/
Processed chlorine-free or PCF indicates that no dioxin-releasing chlorinated compounds were used to bleach the product.

Anyone out there remember Times Beach, Missouri? This tiny little town had their streets sprayed with dioxin to keep the dust down in the summer. The cancer rate rose to astronomical levels, the town had to be evacuated, the city demolished, the roads recovered. It sits alone today near the edge of the river as a park. Have you ever wondered what the run off did to the fish and why they are named whitefish?

✓**Check Green Seal Certified** products that are awarded by a nonprofit group that rigorously evaluates production with standards set for environmental responsibility companies have substantiated their products or service proving it is not harmful to you or the environment.
http://www.greenseal.org/findaproduct/index.cfm

~GREEN BEINGS —SAVE GENES~

Tea party anyone? Not in Boston, at home you can sun-brew your tea. Large sun-tea brew jars are readily available. For a variety of teas to brew, visit recipes dot com. http://allrecipes.com/Recipe/Sun-Brewed-Mint- Tea/Detail.aspx
Better yet, attend a Green Tea party. www.bigteaparty.com. Big Tea Party's newest project, *Green Tea Party--It's Elemental* reflects the group's commitment to inspiring young people to think critically, eat healthier, and adopt a more ecologically friendly lifestyle.

Bottle your own water: If you follow the eight-glasses-of-water-a-day recommendation, it will cost you about .49 cents for an entire year if the water was taken from the tap. Compare this to purchasing bottled water costing $1400 annually. This is one of the easiest ways to save money immediately: stop purchasing bottled water, period. Carry a reusable water bottle or thermos or any refillable earth-friendly bottle. Bottled water is in fact no better and often of lesser quality than municipal water.
http://www.nytimes.com/2007/07/15/weekinreview/15marsh.html?_r=1&oref=slogin
Two giant bottling companies have recently admitted that they use municipal water, with an extra cycle through a charcoal filter. They bottle and package it as if it is pure

~GREEN BEINGS —SAVE GENES~

spring water. Yes, that would be Pepsi's Aquafina and Coca Cola's Dasani and Pure Life brands water are tap water. Not only have the companies misrepresented their products, but also they add 60 million plastic water bottles to the landfills every single day.

The plastic bottles require 20 million barrels of oil to manufacture, transport to the consumers, then onward to the landfill every year. Talk about energy consumption! There are absolutely no health benefits of bottled water. No wonder California has banned this nonsense. According to the United Nations Food and Agriculture Organization (FAO)

- Bottled waters have no greater nutritive value than tap water
- Bottled water is 500-1000 times more expensive than tap water
- By the gallon Evian water costs $22!

The Natural Defense Council found in an independent test of 1000 bottles from 103 brands, that most were good quality. However, one-third of them had chemical or bacterial contaminants exceeding those allowed in big city EPA-regulated tap water. If we only had microscopic vision, we could clearly see that one-third of the manufacturers were tempting our immune systems. For the record, city water must be tested more frequently than bottled water. If you use well water, it should be tested yearly for toxins, pesticides, minerals and

bacteria.
http://assets.panda.org/downloads/bottled_water.pdf

In addition there is may be another problem with bottled water and it is called Bisphenol A (BPA). It is a component of polycarbonate in plastic baby bottles and sports bottles. In April of 2008, the National Toxicology Program raised concerns that exposure to BPA during pregnancy and childhood could impact the developing breast and prostate, hasten puberty, and affect behavior in American children. You see BPA mimics estrogen. Just to be safe, carry your own water in an aluminum-lined or BPA free plastic bottle.

Restore romance: Enjoy a bees wax candle lit dinner at least once a week. Even the children find this a delightful change in routine and it provides another opportunity to teach them about energy conservation, efficiency and, of course, romance. While in the mood, remember to use beeswax or soy candles. Don't purchasing paraffin candles because they are made from crude oil. Use matches to spark your fire. Avoid lighters; they contain petroleum fuel. (See how addicted we are to this stuff!)

Not too romantic: One child only please— There is at least 6.6 billion people living on the planet today. The United Nations predicts the population to swell to nine billion by mid century. They also claim that 54 acres of land is required to sustain one average human being today with food, clothing and other resources extracted from the planet. Continuing such population growth is unsustainable on one planet.

If the world population lived as US citizens, more than five planets would be required to sustain everyone. To learn more about how our consumption habits affect the environment and people around the world, please view this amazing short video by Annie Leonard. After ten years of research her incredible work is revealed in *The History of Stuff*. This is a must see resource.
http://www.storyofstuff.com/
http://www.sciam.com/article.cfm?id=10-solutions-for-climate-change

If you must have seconds, Ms. Piggy: A second refrigerator or freezer should never be put it in the garage. During the winter months the temperature sensor may not be activated if the air temp falls below 42 degrees in the garage. This means the foods in the freezer may actually go through cycles of defrosting and freezing depending on the outdoor temperature.
If you must have a second, keep it in the basement. This definitely saves money in the summer months when it can

~GREEN BEINGS —SAVE GENES~

easily become over 100 degrees in a garage. Keep your freezer full. This helps to decrease the 325 kWh electricity use. Remember that buying in bulk saves on excess packaging, too. Freeze some of the abundant local harvested food for those leaner times.

BYO Cup: No matter if it is a trip to Starbucks or any coffee shop, bring you own coffee cup. Most places offer $.10 discount for bringing your own cup. If not, give up the Styrofoam cups anyway. There is no recycling them. They are another toxic insult to the earth. Try the disappearing continent cup for a reminder of a changing landscape showing sea-level rise around the globe. http://teejayvanslyke.com/?p=109 WSOCtv.com http://www.wsoctv.com/green-pages/10845915/detail.html
http://www.wsoctv.com/green-pages/9755576/detail.html

Choose your chest size (in freezers): A chest freezer is claimed to be about 25 percent more efficient than an upright or side freezer. A ten-year old freezer uses 325 kW of energy and a new freezer has improved only to 291 kW. That means it would take your entire lifetime to recoup the savings each month. This is not really worth replacing.

Also, never add hot food to a freezer or refrigerator. It is always better to wait for food to reach room temperature before storing away. Eco Kids: http://www.ecokids.ca/pub/eco_info/topics/energy/energy_efficient/energy_quiz.swf

Just don't paint me blue: Paint indoor rooms with light colors. Walls that are of a light color reflect 80 percent more light than dark walls. Less electricity is used to light up the room. When painting the outside of your house, keep to a light color if you live in a warm climate, or a dark color if you live in a cold climate for the same reasons of absorbing and reflecting light and heat.

Look for paint that is low volatile organic compounds (VOC). They don't contain the solvents or toxic metals of conventional paint that causes smog and ozone depletion. Paint is a hazardous waste product that needs to be taken to toxic waste sites. Do not place them into the trash. Did you know that the usual household storage of paint is about three gallons? Some paint companies (including industry leaders Dunn-Edwards and Kelly-Moore) sell an array of latex paints recycled from just such residential leftovers and they are approximately 50 percent cheaper than new paint. Recycled paints have old recycled prices but most meet the Master Painters Institute's performance standards. Still more manufacturers now offer VOC paints and they don't leave that paint fume smell! Earth 911: http://earth911.org Eco Kids: http://www.ecokids.ca/pub/eco_info/topics/energy/energy_efficient/energy_quiz.swf

Clean your furnace filter: Clean or replace air filters every month when in use. Energy is lost when air conditioners and hot-air furnaces have to work harder to draw air through dirty filters. Cleaning a dirty air conditioner filter can save five percent of the energy used. That could save 175 pounds of CO_2 per year. http://www.powerscorecard.org/reduce_energy.cfm

Whistle while your work: What you can't whistle? You will get plenty of help with this little gadget that sounds off when the furnace filter is dirty. A clogged filter robs you of money. Because it is so hard to remember when to do all these tasks, some genius out there thought of this device. The furnace filter whistle sounds when your filter becomes 50 percent clogged letting you know it will soon need replacing. Clean filters help to maximize efficiency of heating and air conditioning too.

Make your bed: If you own an energy guzzling waterbed...think about getting rid of it or at least make your bed. A waterbed uses about 100 kWh. By covering the bed one will save energy from heat loss. The covers will insulate it, and save up to 33 percent of the energy it uses. American Council for Energy Efficient Economy: http://www.aceee.org/consumerguide/chklst.htm

Forever in blue jeans: Use old jeans to seal up the largest air leaks in your house. The worst culprits are usually not windows and doors, but utility cut-troughs for pipes ("plumbing penetrations"), gaps around chimneys and recessed lights in insulated ceilings, and unfinished spaces behind cupboards and closets and especially around the inside basement foundation. Pack that area tight with old jeans. The tightly woven fabric makes an excellent air barrier.

~GREEN BEINGS —SAVE GENES~

Multi-tasking at home: Remember: when you must wash the car pull it up onto the lawn. Yes, that water run-off helps to keep the grass green. Don't forget you can use the same biodegradable, phosphate free soap on your car and it won't hurt your lawn. Now that is a green idea!

Dry cleaning chemicals: can be harmful to your health and the health of the environment. Over 85 percent of cleaners are now using perchlorethylene "perc" an inorganic compound. The EPA found that perc is a risk to humans. (It has already made it to the "probable" cause-of-cancer list in California). There are several alternatives. The best is to hand-wash clothes at home. Research has proven even those with a Dry Clean Only label can be safely cleaned at home.

Wet cleaning: Not a freebie but free tip is it won't cost you any more than what you already pay for dry cleaning. This occurs in controlled water, ph and revolutions, time and temperature for the type of fabric. Most stains are easily removed however some require pretreatment with an oil-based solution. The EPA considers this to be the safest of all professional cleaning methods. It is safer for the planet's air and water. For a list of professional cleaners in your state that use Wet Cleaning or Liquid Carbon Dioxide visit http://departments.oxy.edu/uepi/ppc/clearner_near_you.htm http://www.nodryclean.com/map/state.html. Union of Concerned Scientist: http://www.ucsusa.org/publications/greentips/1000-wet-cleaning.html

Recycled carbon dioxide? That nasty greenhouse gas can be trapped from the existing environment and actually put to good use as a CO_2 cleaning agent. This is not harmful. In fact, it is the same form used to carbonate soda.

This is how it is used as a cleaning solvent: The clothing is placed into a $40,000 machine where all air is removed from the chamber. Then pressurized CO_2 gas is the pumped into the chamber. Clothes are soaked in the gas and that effectively removes all dirt and stains. After about ten minutes the liquid form of CO_2 is added to soak up all the debris. Then both the liquid and the gas form of CO_2 is removed and viola clothes are cleaned! The CO_2 is captured as a by-product of existing air space that would otherwise be released into the atmosphere. Since only about two percent of the CO_2 is returned to air with each load of clothing, it's actually helping to decrease the global warming problem. The CO_2 cleaning method also uses less energy than traditional dry cleaning by not using heat and chemical solvents like perc. Remember to take a reusable laundry and garment bag with you. Union of Concerned Scientist:

Silicone cleaning: This is a proprietary technology that employs a silicone-based solvent to clean clothes. The solvent itself is currently considered safe for the environment because it degrades to sand and carbon dioxide, so it has been advertised as "green cleaning". But it has caused cancer in lab animals in EPA studies. In addition, it is manufactured using chlorine, which can generate harmful dioxin emissions.

Some other **cleaning methods advertised as "green"** are not as environmentally benign as they may seem. For example, a solvent called DF-2000 being touted as an "organic" dry cleaning fluid, but it is actually a petroleum product manufactured by ExxonMobil. It is indeed organic in the same way gasoline and perc are organic: it contains a chain of carbon atoms. But the EPA lists DF-2000 as a neurotoxin and a skin and eye irritant for workers and its use can contribute to smog and global warming.

Gratefully these tips come from: US Department of Environmental Protection:
http://www.epa.gov/dfe/pubs/garment

Chapter Seven

Reduce Reuse Recycle

Reduce

Hey thinker: Here is a tricky question! Which requires more energy...a paper towel or an electric dryer to dry your hands? Careful...surprise it is the paper towel. Did you know electric dryers are twice as energy efficient as paper towels? This even holds true when using recycled paper.

It is true the production of electricity that powers the electric dryers generates carbon and therefore greenhouse gases. But, the production of paper towels is twice as harmful to the overall environment. Why? It is because the manufacturing of paper towels results in many pollutants including chlorine being discharged into the air. You know how stinky paper factories can be: that, coupled with the fact that paper towels are made from virgin wood rather than recycled or scrap left over wood product, make it a double whammy to global warming. The loss of the carbon-absorbing trees and the heavy pollution coming from production of paper towels makes an easy choice for consumers when the occasion occurs. The best tip is just to air dry your hands or shake until dry, though that frequently isn't an option. This great tip is compliments of the Sierra Club's Green Life.

What about that new television signal? As of Feb. 17, 2009, analog TV signals will cease being broadcast. But 88 percent of all households get their programs via cable or satellite, so analog TVs in those households can keep plugging along just fine. And even if you don't have cable or satellite, you'll be able to purchase a digital-to-analog converter. This is a much less expensive option than replacing the entire set.
But, if you do purchase a new television remember LCD screens use the least energy (followed by plasma screens), and cathode ray tubes (CRTs) use the most. Here are a few facts about TV energy consumption:

- A plasma TV uses 30 percent more energy than the same size LCD version.

- A CRT TV uses three times more energy than the same size LCD version.

- Standby mode consumes ten to 23 percent of all that juice.

Consider plugging all your electronics into timer-equipped power strips to turn them off completely, at least overnight http://www.sierraclub.org/howgreen/screen/answer.asp

No royal flush here: The Pharmaceutical Drug Disposal Take Back Program has become a necessary sign of the times due to the amount of drugs identified in our water system. Consumers should be aware many drugs are not filtered from the water. Our bays, rivers and streams are contaminated with pharmaceutical drugs that may harm the survival of fish.

In fact antibiotics, anticonvulsants, mood stabilizers and sex hormones have been identified in the drinking water of 41 million Americans according to the Associated Press.
http://www.usatoday.com/news/health/2008-03-09-water_N.htm

Dispose of pharmaceutical waste at hazardous waste take-back sites. Do NOT dispose of unused/expired medications in the trash or down the toilet. Ask for medications with low environmental impact. Fill Rx for only the amount you will need with refills as an option. Encourage your pharmacy to take back unused or expired medication. With a smart diet and exercise many medication will not be required.

http://www.teleosis.org/hn-drugs-in-our-water.php
http://www.whitehousedrugpolicy.gov/pda/022007.html

Be aware of the warning signs: Look for #3 in the triangle-shaped recycle symbol on products representing (polyvinyl chloride) (PVC), number 6 (Polystyrene) and number 7 (Polycarbonates). The Poison Plastic PVC plastic, commonly referred to as vinyl, is one of the most hazardous consumer products ever created. Only Sam Suds could explain it so well. Please see this informative three-minute cartoon video:
http://www.revver.com/video/77761/sam-suds-and-the-case-of-pvc-the-poison-plastic

After viewing you will know why communities living near polyvinyl chloride facilities suffer from groundwater and air pollution. These health risks include angiosarcoma of the liver, lung cancer, brain cancer, lymphomas, leukemia, and liver cirrhosis.

Children can be exposed to phthalates by chewing on polyvinyl toys. While it is still legal for US retailers to sell children's and baby toys containing dangerous PVCs. The European Parliament voted in July 2005 to permanently ban the use of certain toxic phthalates in toys. Reducing these products comes easy once the wool has been pulled from your eyes.

~GREEN BEINGS —SAVE GENES~

In March 2008 US Target stores claimed that they are halting the sales of such plastics in their stores. How do we know when we are exposed to those poisonous toxins? It is easy to identify the odor. Remember the scent of a brand new car or a shower curtain, a new doll? That's the smell of poisonous chemicals releasing gas from the PVC. One of the most common toxic additives is DEHP, a phthalate that is a suspected carcinogen and reproductive toxicant readily found in numerous PVC products.
http://www.coopamerica.org/pubs/realmoney/articles/plastics.cf
http://www.besafenet.com/pvc/about.htm

Plastic bags are close to being banned, too. In the meantime, they should be returned to the grocery stores where they were obtained. The stores will recycle them for you. Unfortunately, they can't be recycled in many recycling centers. They get caught up in the machinery and cause more expense than they are worth.

Reduce a bad idea: Co-op America suggests, that when you find something that just can't be recycled...return it to the manufacturer and tell them to produce or package a product that stops waste! Buy products in bulk when possible to decrease the amount of packaging individual products. It is far better to prevent the production of harmful products than it is to figure out how to recycle or remove it from our air, land and water later. When you buy fresh products, there is no packaging.

Here is another example; the packaging of yogurt and butter tubs is not recyclable at many facilities such as Waste Management. They state that molded plastics require different melting temperatures than poured plastics found in milk jugs or detergent bottles. What if we all returned these yogurt containers to their parent company and asked for something that won't further destroy the environment?

Speaking of waste, think how often your office trashcans are emptied of its plastic bags and contents. Most often the plastic liners and cans are far from full. But when the routine janitorial service schedule roll around a nearly robotic thing happens: even with one cup in the trash the entire bag is pulled out and a new plastic bag replaced.

Here is another opportunity to become part of the solution in preventing further global pollution. Take a look and see what is in your waste can. See if there is a common area for recycling paper. Bring your own cup. Take food waste such as peelings home for your compost container. Be aware of what you are tossing in the trash. Reduce the urge to outperform the basketball stars. Involve your co-workers. This is another great tip from our favorite Green.Life@sierraclub.org

What about Christmas trees? This site offers a zip code locater for a tree recycler near you. In addition you may read about the environmental debate regarding real trees vs. artificial trees. This is a tough call. Consider cutting a healthy tree or bringing more plastic your home? Of course you'd only bring in the plastic once vs. cutting a tree every year. This is another reminder proving that every choice we make has an environmental impact. The best solution is to purchase a rooted evergreen tree. After the holidays either plant it in your yard or donate the tree to your city for a nice little tax deduction before the end of the year. http://www.christmastree.org/home.cfm

Feel like nobility: Little Lord Fauntleroy when you use cloth napkins: not paper. Use washable napkins and towels rather than paper products. Only one-third of paper comes from recycled paper and cardboard in the US paper mills. The remaining two-thirds come from virgin fiber, or freshly cut trees. By using recycled paper products, we are saving valuable trees that filter and clean our air and saves almost $2 billion dollars in

production costs for energy (11.5 billion kW) if all paper used had a 50 percent recycled content.
Please remember to look for a product that states "made from recycled material". Don't buy over-packaged products that waste our resources. Kings County Natural Resources Kids:
http://www.metrokc.gov/dnr/kidsweb/whats_in_garbage.htm

Reuse

Just do it! Make the switch to reusable and rechargeable batteries. When you consider the number of cameras, flashlights, remote controls, and battery-operated toys, there is more than two pounds of packaging waste plus all the toxic material in disposable batteries. Rechargeable may cost a little more to purchase but they save you money in the long haul.
Get rid of your old batteries now at several drop-off disposal sites. Car batteries can also be recycled. They contain over 20 pounds of lead, three pounds of plastic and one gallon of sulfuric acid. Now a little poetry for those chemists out there:

> *Johnny was a chemist's son,*
> *But Johnny is no more.*
> *What he thought was H2O,*
> *Was H2So4 (sulfuric acid).*

Please recycle all batteries. Earth 911: http://www.earth911.org/master.asp?s=ls&serviceid=126 to find your closest drop off location. Or call number at 1 (800) 8-BATTERY or visit www.batteryrecycling.com, Battery Solutions 734-467-9110

Calling all salespeople, resell all your treasures by advertising just about anything to your heart's desire on www.Craigslist.org. There is absolutely no listing fee or hidden charge of any kind. You can find a room to rent, a car to purchase or employment. And, you can do this in many cities and countries around the world. After you log on, scroll down the right side of the screen and pick your location. http://washingtondc.craigslist.org/

Holy sheet! What to do with those sheets and towels or even used clothing filled with holes and no longer useable? Donate them to animal boarding shelters that reuse them for pet bedding. Car washing facilities also appreciate reusing those donations of soft well-worn towels.

Repair and reuse not replace: CDs/DVDs/Game Disks: Send scratched music or computer CDs, DVDs, and PlayStation or Nintendo video game disks to AuralTech for refinishing, and they'll work like new: 888-

454-3223, www.auraltech.com Exercise videos can be swapped at www.videofitness.comwww.videofitness.com

Styrofoam anyone? Here is another place to unload more landfill waste products; UPS stores will accept any flat sheets of Styrofoam for reuse for their product packaging. Small nurseries and mailing stores will too, take back Styrofoam and bubble wrap Still no home found for all that individually shaped Styrofoam.

Diamonds on the soles of the shoes ♪ Faster than weeds: "But I just bought you new shoes!" Here are three great opportunities to turn those used tennis shoes cluttering up your closet into diamonds for someone else! **Perpetual Prosperity Pumps Foundation** pppafrica@mchsi.com is your best option. With every 500 pairs of tennis shoes donated one family is lifted out of poverty. What an easy and thoughtful way to recycle and uplift the lives of so many.

All US collected shoes will be sent to Ghana for the local people. Your old shoes allow families an introduction to the MORE system, which stands for Modular Organic Regenerative Environment. After completing the training program, each family receives hands-on training in raising crops and animals. The program includes access to water and pump system, seeds, fruit trees and livestock. Yields are ten times greater than the traditional African agricultural methods. Families can generate more than 40,000 pounds of food each year. This program will definitely benefit more people than any shoe-collection program out there. Please visit the web site and learn how easy it is for your place of employment, worship or education to collect used shoes and be a part of this fantastic endeavor.

Other options include: *One World Running* will send still wearable shoes to athletes in Africa, Latin America or Haiti. Contact www.oneworldrunning.com
Sports equipment can be donated or in some cases sold at this local outlet. Play It Again Sports stores outlet, 800/476-9249, www.playitagainsports.com.

If the shoes are badly worn, Nike's Reuse a Shoe program will turn old shoes into playground equipment or flooring. www.nikereuseashoe.com another donation program, http://www.recycledrunners.com

I see the light: Trouble seeing? Get free glasses at your local Lion's Club. And donate any unused prescription glasses, too. Another resource to find homeless glasses is in dry cleaners across America. Most have a basket of lost and forgotten glasses to glance through.

Pretty darn smart: With your great efficiency skills and memory all the books of the world, perhaps even this one can be read for FREE through the public library. Yes, you are so smart. Some libraries even loan artwork for free and table top sculptures too on a monthly basis. Most have intranet computer services. They are still absolutely free and are the compliments of your tax dollars. If you prefer owning your book, you can swap paperbacks with other conservation minded people at www.paperbackswap.com

Yesterday's news: There are dozens of uses for something black and white and read all over. You got it. Yesterday's newspaper provides more than news. This site lists 20 common uses, a few of them are listed below-but you can be creative.
www.apartmenttherapy.com/la/look/20-household-uses-for-newspaper-once-youve-finished-the-crossword-puzzle. 050121

Gift-wrap
Packing material
Cleaning windows
House training a pet
Cage liners

~GREEN BEINGS —SAVE GENES~

Weed block in the garden
Germinating seeds
Maintaining shape in hats, boots, purses
Drop cloth while painting
Paper Mache toys or fruit
Table protector for messing meals
Under the high chair
Prevents ice and snow from auto windshield
Moisture barrier for wet shoes
Odor eater in the refrigerator or plastic dishes
Starting a fire and making a thinking cap

Mother knows best: Here is another remnant of the Great Depression and a true money saver tip. For all those old greeting cards you have tucked away, there is a new use. When you need a package To: and From: label, just cut a funny rectangular, square or circle shape from the old card, keep the fold in the new label and attach to the gift.

Or, for fancy tree-saving greeting cards, business cards and stationary visit Tug Boat Paper http://www.tugboatpaper.com/site/welcome.php. They produce tree-free hand-made paper products with soy-based ink. Now that is Environmental Consciousness!

Use canvas or reusable grocery bags: Some stores offer a three to five cent refund for reusable bags you fill with groceries and many will be charging $.15 cents for each plastic bag you use. *Bring That Bag Back* is a slogan currently traveling through the US grocery markets. This idea is pretty simple. Instead of choosing paper or plastic at the grocery store, bring your own reusable bags.

"I would have been happy to give it to you but you forgot to bring your own bag"

Plastic bags including trash, grocery and sandwich bags are made from petroleum. There are over one trillion plastic bags used on the planet each year. The US uses 110 billion plastic bags that are expected to take 1,000 years to decompose in the landfills. How can any substance be expected to decompose without air, water, and light? The most common type of plastic shopping bag is the supermarket type. It is made from a man-made polymer called polyethylene. Unfortunately, microorganisms don't recognize the substance as food and will avoid contact with plastic bags.

And if you think paper is better...re-think. Did you know it requires 14 million trees to produce ten billion paper bags? And of course the paper comes from our forest trees that are the major absorbers of the greenhouse gasses. (Remember: Plant a Tree) So, cutting them down for bags is one insult. The second problem is the high heat, pressure and chemical solution used by stinky paper mills to produce the paper bags. This adds to global warming.

The third insult is the toxic chemicals that are added to the air. This pollution produces acid rain, more asthma (in children and adult onset asthma), and eventually the sediment settles to pollute the waterways. It is a nonstop train-wreck. Yep, the fish get contaminated and the people eat the fish and we all get sick from not carrying our reusable bag to all stores.

The end of this story is using <u>reusable bags</u> and <u>containers</u> are the solution. Don't forget, if you order take-out food; remind them at the time of the order that you will bring your own container too.

http://www.reusablebags.com/facts.php?id=7Com:
http://www.reusablebags.com/facts.php?id=7
http://www.wsoctv.com/green-pages/10845915/detail.html

Recycle

Recycling is near the top for energy and resource conservation. The most common recyclable items are glass, paper, aluminum cans, plastic, and cardboard but every municipality has specific abilities and unique collections, for example bicycles. Check with your local public works department. According to the EPA the US recycles approximately 32 percent of its waste. That reduces greenhouse gases to an equivalent of removing 39,618 cars from the road. Increasing the recycling rate just three percent more to 35 percent would reduce greenhouse gas emissions by an additional five million metric tons of CO_2 equivalent. Save on the cost of trash pickup, trash bags and trash cans too because recycling is a free service and most cities will provide the recycle container.

Do you know that Waste Management System will pay your city for its trash? An even more interesting, is that a tractor-trailer load of crushed aluminum is sold for a whopping $35,000! Almost as interesting is one recycled can provides enough energy to power a television for 30 minutes! There is value in your trash...waste not, want not. To see what is in the typical trash can check out this site. Keep a **Global Basket** in your kitchen. Designate it as your recycling container. This is a handy way to shuttle your items to your outdoor recyclable trashbin or collection center. Kings County Natural Resources: www.globalbasketcase.com http://www.metrokc.gov/dnr/kidsweb/whats_in_garbage.htm

More to recycle? Call Earth 911. Their site offers information for recycling just about any product. It is easy just enter your zip code and you will be directed to the closest facility.
Earth 911: http://eart911.org/outreach/public-service-announcements

Electric appliances: This is a tough thing to get rid of. You know the guy who has two refrigerators on his front porch? Let him know about The Steel Recycling Institute or visit their web site at www.recycle-steel.org. If they are still in working condition, Goodwill Industries will sometime accept them as a donation. Worn out and worthless ones may even be accepted. Call Steel Recycling Institute to recycle them. 1-800-876-7274 X201.

Flat and dull again? Buy a

recycled plastic toothbrush or razor from Recycline, and the company will take it back again to be recycled into plastic lumber. Just call 888-354-7296 or visit www.recycline.com

What about that old television? Did you know there were more than 45 million new television sold around the world and 30 million of them were in the US? You can imagine what happened to most of the old ones. As new technology becomes available for digital, flat screen and high definition, there will be even more of the old behemoths on their trip to the local landfill.
But wait. You can help. Less than 12 percent of people are recycling their electronics. Manufacturers must help out in this environmental disaster. Sony Corporation is leading the way in the US with their innovative TV-Take-Back Program. Please support them as they give us a great opportunity to prevent further pollution.
http://takebackmytv.com/page/speakout/TakeBackMyTV
In addition, for the first time, TVs will have to meet lower power requirements for the "on" position or active states to bear the *Energy Star* logo. Currently, Energy Star-compliant TVs must consume less than 1 watt of power in standby mode, or what most of us know as "off". The new specifications, however, include both standby power and active power requirements, and will go into effect in November 2008.
http://www.greentechnolog.com.

No need to Ride your 🚲 **or** 🚴 **to Recycle with Freecycle:** (Say that three times while telling as many friends) The Freecycle Network™ is made up of many individual groups across the globe. It's a grassroots and entirely nonprofit movement of people who are giving (and getting) stuff for free in their own towns. http://www.freecycle.org. This is a great alternative to landfill waste. http://throwplace.com/ In California visit http://ireuse.com/

Going postal: The US Postal Service has started a new program to help customers recycle small items such as electronics and inkjet cartridges. The "Mail Back" program helps consumers make more Earth friendly choices, making it easier to discard used or obsolete small electronics in an environmentally responsible way. Customers use free envelopes found in 1,500 Post Offices to mail back inkjet cartridges, PDAs, Blackberries, digital cameras, iPods and MP3 players – without having to pay for postage. http://www.usps.com/communications/newsroom/2008/pr08_028.htm

Don't forget those old ink cartridges are worth $1 for each cartridge recycled at http://www.recycleplace.com

and $3 at Staples. Reminder, it takes more than ten centuries for a printer cartridge to decompose.

Adios: Let's give all our old cells a proper burial. This gives a whole new meaning to apoptosis (cell death). Most have heard of mercury and lead poisoning too. But did you know cadmium compounds accumulate in the kidneys after being absorbed through the air? The half-life is 30 years. Please don't toss these toxins in the landfills. They will reach the water and the air and eventually you, too. They can be used to refurbish your next telephone or provide it to someone less advantaged or for export to a developing country. Call 770-856-9021 or visit their web site above. Other options include;; www.donateaphone.com or www.reclamere.com 814-386-2927 or www.eco-cell.org. Most cities now have several computer-recycling programs too. Organize a neighborhood computer clean out party and call: Turtle Wings, 410-975-9046 http://www.turtlewings.com/

Hang it up: Not only cell phones, computers, monitors, and electronics too. There are many companies wanting to recycle this electronic trash. Most electronics are filled with components that are valuable to recycle and harmful if left to rot in the environment. Americans toss out more than 100 million cell phones every year. Keep their toxic ingredients (including lead, mercury, cadmium and arsenic) out of landfills by recycling your old cell through Collective Good Mobile Phone Recycling: www.computertakeback.com http://collectivegood.com/

You can even get a tax break by determining the value of the cell phone based on depreciation since purchasing. That money can be donated to a variety of charities from Animal Humane Society to Zest about any non-profit organization you can imagine.

Another company offering to recycle your Technotrash is Green Disk. For a fee of $30 they will send you a special box you may return it to them with 70 pounds of electronic gear. 800/305-GREENDISK, www.greendisk.com. Or, visit Earth 911: http://earth911.org/outreach/public-service-announcements or visit Lamp Recycle Dot Org:
Now a little bit about recycling that hot topic, compact fluorescent light bulbs (CFLs) but lots more on this subject coming up in the next chapter. WWW.LAMPRECYCLE.ORG or email lamprecycle@nema.org. Environmental Defense: http://www.environmentaldefense.org/page.cfm?tagid=609 for information on disposing of these long lasting bulbs. Or, should they burn out before their five-year guarantee, take it back to place of purchase. Another alternative is to take them to a local Ikea store for recycling. www.ikea.com

When the rubber is done hitting the road: Don't forget to recycle those old tires. They can be reformed into dozens of products according to the American Rubber Technology Association. Here are some examples of recycled products to consider for upcoming projects: Rubber Stuff Landscape Mulch, benches, picnic tables, trashcan guards and borders, jar openers, portfolios, 3-ring binder covers key rings too. Join the solution! http://www.americanrubber.com/rubberstuff Perma-Flex: http://www.perma-flex1.com/ equestrian arena rubber footing playground safety surfaces, Rebound Soil Amendment: http://www.americanrubber.com/rebnd.htm Recycled tires through American Rubber Technologies:

~GREEN BEINGS —SAVE GENES~

http://www.americanrubber.com/cgi-bin/shop/store.cgi?action=search&uid=1&category=Accessories

Motor oil revisited: Used Motor Oil Hotlines for each state: 202/682-8000, www.recycleoil.org.

Cardboard boxes: Cardboard is accepted at most recycling centers that recycle paper. We are recycling fifty percent more boxes than just ten years ago. Currently more than 75 percent of all boxes find their way to a recycling center. For instance, Waste Management and Recycle America recycle 65 tons of trash in one hour at their Elkridge Recovery Facility in Maryland. They love cardboard, telephone books, catalogs, office paper and just about anything paper. They process 20,000 tons every month and 70 percent of it is paper product. Visit www.recycleamerica.com or www.UsedCardboardBoxes.com http://www.newsweek.com/id/69435

Tree hugger's No. 1 bio-thinking award product: How about *Cardboard Coffins*...hmmm may want to hold off on that one while you consider this wonderful recycling tip: Donate your organs (no pianos) before you go. Truly, over 96,000 people are on the waiting list for a little part of you. And there are only 6,000 of you out there willing to leave a small piece behind. Oh, don't be afraid. It won't hurt a bit. Call or contact http://www.organdonor.gov/ or contact the Department of Health and Human Services. They will mail you an Organ/Tissue Donor card for FREE. Mention Inventory Code: HRS 00259.

Another amazing recycling effort recently announced on public radio was from a crematory claiming it is recovering the heat generated from their services to heat their building! What a burning desire to be efficient. It warms ones heart.

Fuel for the fireplace: (I confess to you my brothers and sisters, my personal guilt, passion and sin). If you enjoy the warmth and glow of fire, try using manufactured green logs. Seems just about anything that has removed the petroleum from their products is now claiming to be green. Such is the case for Duraflame logs. They offer recycled fire logs for an easy efficient fire. http://simplefire.com/

Earth friendly logs that spare landfills of wax coated cardboard is reshaped into clean burning logs and available through http://cleanflamelog.com/ (Thanks to the Java-log company for this resource). These logs were first produced in 1998 as an experiment mixing dried coffee grounds with wax. They burned so well that a Java Log company was formed. Talk about recycling, this takes the cake and coffee too.

Chapter Eight

Good Ole' Uncle Sam

An ounce of prevention is worth a pound of cure. Slowing pollution is not the answer. We must PREVENT pollution and this saves the environment more than attempts at restoring Mother Nature.

CLEANING UP THE ENVIRONMENT IS NOT THE ONLY SOLUTION

Free home energy audit: You can get instructions from Department of Energy to do your own home energy audit. http://www.eere.energy.gov/consumer/your_home/energy_audits/index.cfm/mytopic=11170
Most energy audits prevent an average of 11, 273 pounds of carbon per year from being dumped into the

environment. The free for homeowners' guide include how to do-it-yourself from Energy Star.

Call Maxwell Smart: For a few hundred dollars, an expert can come to your home and do a complete inspection and prescribe the recommended changes to improve your utility bills. Try to find a company that offers money back guarantees. *If you don't save the cost of the inspection the first year, your money will be promptly refunded.* Several companies bank on the idea that most people will not follow the recommended tasks. They just don't realize how smart you are. Those bargains are definitely out there. If they don't offer it, ask for it in writing. Energy audits create local jobs that won't be outsourced. So you are helping yourself and your local community by hiring a professional.

Free tax credits and incentives: Energy conservation through heating, cooling, and water heating equipment all offer plenty of rewards for your effort. Purchasers of qualified energy efficient appliances are eligible for tax credits up to the total expenditures on such items. The credit can also be applied to labor costs for assembly and original installation on this property. Eligible property and maximum credit amounts are as follows: but look for new bigger and better tax cuts after November 2008.

- Electric heat pump water heaters $150 to $300
- Advanced main air circulating fans ($50)
- Natural gas, propane, oil furnace or hot water boilers ($150)
- Geothermal heat pumps ($300)
- Central air conditioners ($300)
- Natural gas, propane, or oil water heaters ($300)
- Electric heat pumps ($300)

There is currently a bill waiting for approval to increase solar tax incentives to $10,000. Write your congressman and remind him or her of your voting capability and how you want him to represent your needs. US Department of Energy: http://www.energystar.gov/index.cfm?c=Products.pr_tax_credits

More help is on the way: Take advantage of federal income tax credits that reduce what you owe to Uncle Sam or increase your tax refund by $250 to $3,400 for purchases of hybrid-electric or diesel vehicles. Amounts are based on the vehicle's efficiency and fuel savings. Details in English and Spanish are on the Alliance's website – Alliance to Save Energy: www.ase.org/taxcredits

House of Representatives passed H.R. 5351, the "Renewable Energy and Energy Conservation Tax Act of 2008" by a vote of 236-182. The tax package reinstates the credit for homeowners who make certain energy efficiency improvements to their homes and extends it until December 31, 2009.

No rocket science here: Knock down your electric bill with DSIRE: DSIRE is *Database of State Incentives for Renewables and Efficiency.* http://www.dsireusa.org DSIRE offers a comprehensive source of information for every single state, local, utility, and federal incentive that promotes renewable energy and energy efficiency. Choose one or both databases to search. The tax incentive closes in on the gap and saves tons of carbon from being dumped into our air. Interstate Renewable Energy Council: http://www.irecusa.org/index.php?id=24

His and hers: Home Energy Rating System is a scoring system from zero-100. A lower energy rating number reflects a more efficient home. Each point is equal to one percent energy savings. This is a useful scale for preconstruction estimates of energy saving tactics. http://www.energystar.gov/index.cfm?c=bldrs_lenders_raters.nh_HERS

FREE weatherization assistance program: This program serves low-income families free of charge and limits,

according to federal rules, the amount of money that can be spent on any single residence. The average expenditure is $2,744. Please visit the Dept. of Energy web page to check eligibility.
http://www.eere.energy.gov/weatherization/apply.html

VOTE: Support your politicians who are out to help you: Know your congressmen and delegates. Call their office and remind them that you will be voting on their actions. Here is something to follow closely: Speaker Nancy Pelosi is an advocate of a renewable-energy bill. It includes the extension of tax credits through 2011 for facilities that adopt many renewable-energy systems at an estimated cost of $6 billion dollars over ten- years. Solar energy and fuel-cell energy use would qualify for a 30 percent investment tax credit through 2016. The bill also includes tax credits for cellulosic ethanol, biodiesel production, plug-in hybrid vehicles and energy-efficient improvements to homes and businesses. The bill is expected to pass the House but faces an uncertain fate in the Senate. Call your Senator!!!

Strange but true and free solar panels: on your home for **Free** from Citizenre Join the Solution. Not only that, you can earn two percent credit on every home energy bill that you refer to the program. Now

that is a pointed idea. Visit:
http://www.citizenre.com/web/index.php

Break the Piggy Bank: Are you wondering where you should place all the $$$ you saved in Chapter One? Try investing in environmentally conscious mutual funds. You can find them at www.socialinvest.org

Now is the time for all good men and women: Pull out the latest utility bill and admire your achievements. Fill in the numbers on your latest utility bill here then pat yourself on the back. Grab some of those rescued dollars and enter Phase Two. Good job! You are definitely part of the solution!

- Month:

- Utility:

- Costs:

- Kilowatts Used:

http://greenerchoices.org/globalwarmingathome.cfm
http://www.coopamerica.org/pubs/greenpages/greentips/january.cfm?notext=t
http://www.insidethebottle.org/california-source-water-would-be-clear-under-new-law

http://www.state.mn.us/mn/externalDoc/Commerce/Appliances__Electronics_110802035138_Appliances.pdf

The Power is in Your Hands
www.powerisinyourhands.org. A new web site from the Alliance and 20%20 partners with extensive tips and resources to arm consumers with the power to manage their energy bills this winter.

Free Alliance to Save Energy resources. Obtain a free booklet, recently updated and redesigned in 2005, *Power$mart: The Power is in Your Hands,* by calling 1-888-878-3256 or previewing a PDF version. An interactive *Home Energy Checkup* allows handymen (or women) to troubleshoot their homes' energy waste while calculating efficient improvements.

Free Department of Energy resources. Obtain a free booklet, *Energy Savers: Tips on Saving Energy and Money at Home,* in English or Spanish by calling 1-877-337-3463 or view an online PDF version.

Free Environmental Protection Agency resources. Obtain a free copy of *Guide to Energy-Efficient Cooling and Heating,* which is available at www.energystar.gov from the heating and cooling product pages or by calling 1-888-STAR-YES (1-888-782-7937). Download the ENERGY STAR Action Guide *5 Steps You Can Take to Reduce Air Pollution.*

Free home insulation booklet from North American Insulation.

Free booklet for New Yorkers: *It's Right... and Right at Home*, a brochure with energy saving tips. Contact 877.NY-SMART (877.697.6278) or residential@nyserda.org.
www.GetEnergySmart.org, the consumer web site of the New York State Energy Research and Development Authority (NYSERDA).
Sources: Dept. of Energy, Environmental Protection Agency, and Alliance to Save Energy

PHASE TWO

Chapter Nine

Small Investments with Big Payoffs!

A little goes a long way...a dollar spent on energy efficiency would save seven times more carbon dioxide than a dollar spent on nuclear power. (Sierra Magazine by Paul Rauber *Why Not Nukes*) January 2007. Other replicated studies have proven $1 spent on energy efficiency will save the consumer $4 worth of expenditures. To consume or not to consume, even utilities agree a little energy efficiency *goes a long way*.

Welcome to Phase Two. Here, you can start spending a little to save even more. Start your purchases with the least expensive item and work your way up to even more environmental efficiency. You're on the way now. Enjoy!

Abe to

Ulysses

~GREEN BEINGS —SAVE GENES~

Three Cent tip: wise guys, choose wisely and drink responsibly: As you congratulate yourself for completing Phase One keep on popping those tops on your wine bottles. Did you know "cork" comes from an incredibly diverse and endangered landscape? You can help protect the nearly three million hectares in the Mediterranean from impending destruction due to the recent influx of plastic screw bottle tops. Manufacturers' are choosing to use plastic tops because they cost three cents less than a cork bottle top. Don't fall for plastic trap. Spend just three cents more to protect the environment.

A Cork-Oak tree lives for 300 years and gives up renewable bark every nine to twelve years for the production of cork bottle stoppers. Not one tree is cut down. This farming provides income for more than 100,000 people and home to a diverse bird and animal population as well as several endangered species. Don't let all those critters get cork screwed? So pull out that bottle and corkscrew raise your glass for a little cheer…just remember to take it from a Corked Oak bottle. World Wild Life Fund: A corked bottle of wine cost only pennies more than a *screwed up* oops, a screwed on plastic top and well worth the price.
http://assets.panda.org/downloads/cork_rev12_print.pdf

Be sure to save those corks too. They can be made into cushy floor tiles. What a charming addition to your wine cellar! Contact Yemm and Hart from the Show Me State of Missouri for their cork recycling collection as well as their fabulous cork flooring. P.S. you can even get your three cents and more back for your corks. This is a great way for schools or groups to make easy money. www.yemmhart.com/news+/winecorkrecycling.htm

Remember, "Take care of the pennies and the dollars will take care of themselves". Pennies are now worth three cents in copper value alone. Hold onto them. If the government decides to remove pennies from the currency, they can be legally sold for the copper. That is an immediate 200 percent profit. You can't get that kind of return in the bank. Get that old piggy bank out and put it to a use, squeak, squeak.

Purchase recycled paper products: Paper products come from trees. If every household purchased just one package of recycled paper napkins, one recycled box of tissue, one recycled roll of toilet paper and one roll of recycled paper towels how many trees would be saved in one year?
a. 100,000 trees?
b. 500,000 trees?
c. 1,000,000 trees
d. 2,130,900 trees?

~GREEN BEINGS —SAVE GENES~

If you picked D you are correct. We need trees to absorb CO_2 generated by coal fired power plants and automobile exhaust. National Resource Defense Council: http://www.nrdc.org/land/forests/gtissue.asp

Purchase only paper products that have a high-recycled content. Check the amount of post consumer content. Post consumer refers to paper that would have otherwise been dumped in a landfill. Always purchase unbleached or color dyed paper. Be mindful of the toxicity of chlorine and potential harm to you, your family and all the species in the world. Labels will use these initials:
- Totally chlorine-free (TCF)
- Processed chlorine-free (PCF)
- Elemental chlorine-free (ECF) may be acceptable.

Seventh Generation Company markets 100 percent recycled paper for production of new toilet paper at **83 cents** a roll. In addition to the trees that will be spared, we are also preventing further landfill development, garbage trucks on the road, saving water and preventing pollution. Recycled paper is another win-win. Other brands to support are: Marcal, Planet, Green Forest, Fiesta, Earth First, Atlantic, Pert. Companies to avoid: Cottonelle, Bounty, Scott, Viva, Kleenex, Soft Pac and Charmin.

Who is the airhead? Place aerators on faucets and replace older faucet aerators with new ones that are rated at one and a half to two gallons per minute (GPM), or less. Aerators cost only a few dollars. You can purchase an aerator for about **$1.99** to increase spray velocity while reducing splash and decreasing water consumption. This is an inexpensive way to conserve energy. The US Environmental Protection Agency makes it easy for Americans to be sure products will save water and protect the environment through a partnership program with WaterSense. Look for the WaterSense label on products to be certain. This tip from http://epa.gov/watersense/index.htm and Eco Kids: http://www.ecokids.ca/pub/eco_info/topics/energy/energy_efficient/energy_quiz.swf

Tell your friends: When it is time to send out those holiday cards, at **$2** each you can spread the word about recycled paper. Dolphin Blue sells holiday cards printed with soy ink on paper made from 50 percent PCW recycled and 50 percent recovered cotton ($64.36 for a box of 32; www.dolphinblue.com; 800-932-7715). Dolphin Blue also offers a service to individuals and small businesses in which letterhead and envelopes, bearing your logo or other information, can be printed onto a variety of environmental papers. Another terrific web based company offering the widest range of eco friendly products on recycled paper is Green Printer. Visit their informative site at www.greenprinteronline.com/index.html their slogan; *Save Trees - Print Green*. They print those hard to find (on recycled paper) products such as business cards, bookmarks, calendars, cards, brochures and too much to mention here.

Thar she blows: The National Oceanic and Atmospheric Administration (NOAA) have forecast another active hurricane season.

- Abnormally hot days and nights, along with heat waves, are very likely to become more common.

- Droughts are likely to become more frequent and severe in some regions.

- Hurricanes will likely have increased precipitation and wind.

Be prepared. Toss out that electric can opener once and for all times and use the old fashion **$2.99** priced hand-cranked can opener. Prepare a kit with matches, solar rechargeable batteries,, home-canned foods, and home-bottled water. Don't forget the soy or beeswax candles.

Looks can be deceiving: Philips Energy is offering a light bulb that has the same size and shape of the old-fashioned incandescent bulb. If you are unhappy with the new CFL's, try the ALTO® Lamp. It is claimed to require 70 percent less energy than an incandescent bulb and provide the same soft white light lasting up to six

years. The price ranges from **$3-$6** a bulb.
http://www.nam.lighting.philips.com/us/consumer/energysaver/energysaver.php?mode=2

Install light sensors to your outdoor lights: It happens to all of us, the old memory cells fade. There is a **$5.98** solution. Install light detectors either indoors or outdoors for those occasions where you may need light without having access to the switch. For example, a porch light will come on automatically for your evening arrival home. No need to reset or fumble with a timer. Timers usually need to be plugged into a socket and then the lamp fixture is plugged into the timer. That becomes impossible with a ceiling fixture. This inexpensive little gadget fits between the light bulb and the socket. A small light sensor protrudes from the device. There is no expensive wiring to do. A light sensor is different from a motion detector. Most outdoor motion detectors that add light do so for security and surveillance reasons and they cost much more. Using a light sensor will save lots of time, and energy: yours and the planets. It also will save about $12 and 200 pounds of CO_2 per light fixture. Buy several and save more. Try to remember this great tip...file it in the frontal lobe just under the skull.

Eat quickly with biodegradable compostable utensils: Just kidding, about eating quickly. They won't decompose that fast. They are actually made from heat resistant corn resin and can tolerate your entire meal before melting away. A package containing 20 pieces or more,

costs **$5.99**. You can find them at Trader Joe's or as usual, on the web.
http://www.ecowise.com/product_info.php?cPath=22_187_195&products_id=536

Don't be bi-sacksual with paper or plastic: Go all the way with an eco-friendly reusable grocery bag. You can purchase the grocery store's token dollar bag that doesn't quite hold a gallon of milk, or purchase a real brown bag sized reusable recycled bag in any color you'd like. They retail for about **$6** a bag.

For $7: You can buy one reusable copper coffee filter. This is yet another easy way to decrease our demand for paper products. Most paper products come directly from harvested trees not scrap wood or excess lumber.

Did you know one of the main product produced from Giant Redwood trees is pencil wood? That's right. When those majestic trees hit the ground they shatter so much that their main use is designated for pencils and match sticks. Use a pen or reusable pencil with replacement lead.

Check your temperature and call me in the morning: For another **$7** bucks you can purchase a mercury free digital thermometer instead of using the old fashion shake-it-down kind. Did you know the mercury in those old thermometers is enough to contaminate a twenty-acre lake? This would result in more fish warnings for pregnant women and children. Go ask Alice-from Wonderland and her contact with the Mad Hatter. Did you know the production of stiff brimmed top hats at the turn of the century-contained mercury? The people that produced the hats had a bad reputation for being angry or *Mad as a Hatter*. This is just another little reminder for you to avoid mercury exposure. http://www.grist.org/news/counter/2000/02/16/falling/

Well, well, American ingenuity: Preserve Cutlery has finally come up with a use for all those empty yogurt cups. They too, offer disposable cutlery made from 100 percent recycled plastic. The cutlery is dishwasher safe and can be used again and again before returning them to the recycling center. A 24-piece set retails for **$7.95**.

Double duty: This handy **$8** bag works for you twice and the environment forever. It serves as a laundry bag for you to tote your clothing to the cleaners, not dry cleaners but *wet cleaners*. Leave the bag with the cleaners and when your clothes are ready for pick up the bag is folded inside out and becomes a reusable garment bag for those freshly cleaned products. No more plastic bags. It is pretty handy.

Reserve Preserve products: If you must purchase disposable paper plates and cups visit Preserve for 100 percent recycled and reusable products. They also manufacture personal razors for shaving. To really conserve...grow a beard...only the gentlemen please. Toothbrushes and razors can be purchased too, costing **$5-$11.** The good news is they will take your old ones right back for another recyclable product. Way to go! http://www.recycline.com/index.html

The good, the bad and the creepy crawly: Drainbo's drain cleaner is non-toxic with environmentally friendly bugs that make a quick meal of grease, oil, hair and all that yucky stuff that gets stuck in your tub or sink drain. Billions of micro-biotic critters can be placed in all kinds of toilets, portable or otherwise and into any holding tank too. Gobble-gobble-gobble and that is not coming from a turkey. *Drainbo costs $8 per quart.* www.drainbo.com

Every brain surgeon knows "two heads are better than one": In fact, install two or more low-flow showerheads in your bathrooms to keep up with the old doc. Cost can be as low as **$10** each. If half a million households installed just one low flow showerhead, over 200 million gallons of fresh water would be conserved each day. That is enough to provide fresh water to 3,000 homes for one year. It's easy to save while you shower.

Your best choice in low-flow showerheads is one with a temporary shut-off button that allows you to turn off the water while lathering up or shaving without having to readjust the temperature when you turn the water back on. Not only do the showerheads deliver 1.5 liters of water a minute, they do so while feeling like a ten liters per minute showerhead. At the same time, they can reduce carbon dioxide by 302 pounds per year largely due to the fact that less water has to be heated for your shower. This information is from Seattle City Light in partnership with Puget Sound Energy and Cascade Water http://www.savingwater.org/docs/Showerheads

Hats off to Staples: It is the first major retailer now accepting all types of electronic products for recycling. In addition, Staples Office Supply Stores claim to have saved one and a half million trees in 2005 due to marketing recycled computer paper. Support **Staples'** green efforts: the retailer now sells recycled copy paper, manila envelopes, legal pads and sticky notes, including the Earth wise 100 percent recycled/30 percent PCW file folders. Save a tree, purchase recycled paper from Staples while dropping off old electronic products.

Kudos to Staples is offered with some reservation. We found out-after sending a friend with more than 30 computers to recycle they charge **$10** per computer. Keep this in mind before you offer to take several with you. Contact Staples at www.staples.com

Let sunny lead the way with solar landscape lights: What an evening delight to find your way with lights powered by sunshine captured during the day. These remarkable fixtures can be purchased as low as **$9-$15** each. The best buy is purchasing those with super energy efficient light emitting diode (LED) bulbs. This is a one-time purchase that just keeps renewing itself for you.

Charge me up, Scottie: Rechargeable batteries are the way to go. Disposable batteries contain toxic chemicals, and manufacturing them takes about fifty times as much energy as the batteries produce. For everything you have always wanted to know about batteries visit the Department of Justice for this extensive report. Cost average **$11- $15** for a four pack. http://www.greenbatteries.com

http://site.greenbatteries.com/documents/Battery_Guide.pdf By purchasing solar chargers, there is an added environmental benefit.

Holiday Christmas lights: Replace your incandescent holiday lights with LED holiday lights. Although LED lights may initially cost a few dollars more, they can reduce energy use by more than 90 percent, often paying for themselves in a single season. LED holiday lights will burn more brightly than traditional lights and in excess of 100,000 hours (more than 11 years of use, 24 hours a day). Holiday LED lights are available at: Sam's, Costco, Target and Lowe's. They range from **$10-$18** a strand of 35-120 bulbs. Resource Pacific Gas And Electric: http://pge.com/res/holiday_lighting.html

Now you may ask what I can do with all those twisted and tangled strands of incandescent holiday lights. HolidayLEDs.com is accepting incandescent holiday lights for recycling during the month of January.

How many brain cells does it take to replace a light bulb? Let's put some light on the subject: Incandescent, compact fluorescent bulbs, fluorescent tubes and LED bulbs, wow, wondering what to choose? It seems the best alternatives use the latest technology. Let's take a look.

The traditional incandescent bulbs are actually just small heaters that produce light. They waste a lot of energy making heat. Remember 85 percent of the energy is heat and only 15 percent is from light. The original light bulbs were invented in 1880's. We have made progress in almost every product since then (except the safety pin). Those old bulbs are so inefficient they have already been banned in Australia. If you replace just five light bulbs, you will spare the atmosphere 500 pounds of carbon each year.

Bright Ideas

Compact fluorescent light bulbs: CFL bulbs are four times more energy efficient than incandescent bulbs and provide the same lighting. Costing **$6-$15** per bulb, the cost can add up quickly. Remember the savings will be reflected in your utility bill. Generally, for each bulb you exchange for a CFL, you can expect to save $10 per year and about 140 pounds of CO_2 per bulb. A rule of thumb is to purchase a CFL bulb whose wattage is about one-quarter of the incandescent you're replacing. For example, a CFL bulb in the 15-watt range replaces a 60-watt incandescent.

Do you want more information on watts and lumens? Visit Environmental Defense at http://www.environmentaldefense.org/page.cfm?tagid=630

In order for a CFL bulb to work in a dimmer, it must be specially designed. Read the package before you buy a CFL bulb for a dimmer-controlled fixture.

If every household replaced just five 60-watt incandescent bulbs with CFL bulbs, the pollution savings would be like taking three and a half million cars off the road! And the carbon savings would equal more than 500 pounds per year.

Oh, I hear an objection from the skeptics. Yes, it is true there is mercury in these bulbs. Pull out the microscope. It is four milligrams. If you can remember the old fashion mercury thermometers, they contained 500-3000 milligrams of mercury (remember enough to contaminate a twenty acre lake?). According to the Environmental Protection Agency, a power plant will emit ten milligrams of mercury to produce electricity over the lifetime of an incandescent light bulb compared to less than two and a half milligrams over the lifetime of the CFL. So, by using these bulbs one actually helps prevent mercury from being released into the air from coal-fired power plants. What a coincidence that the things that are so much better for the Earth also prove to be beneficial to your savings account.

This info comes from US Department of Energy.
http://www.energystar.gov/ia/partners/promotions/change_light/downloads/Fact_Sheet_Mercury.pdf
Fast Company Dot Com:
http://www.fastcompany.com/magazine/108/open_lightbulbs.html . Fluorescent tubes are more efficient than compact fluorescent bulbs.
Eco Kids:
http://www.ecokids.ca/pub/eco_info/topics/energy/energy_efficient/energy_quiz.swf Environmental Defense: http://www.fightglobalwarming.com/page.cfm?tagID=608

Times, they are a changin': Today we are recommending CFL's. Tomorrow it will be LED's. And just around the corner is Silicon Valley's Luxim. They have developed a light bulb the size of the candy Tic-Tac that gives off as much light as a streetlight. For comparison an incandescent light bulb gives off 15 lumens of light per watt. An LED will give 70 lumens of light per watt. And this newly discovery Luxim will give off 140 lumens of light per watt. Where there is a will, there is a way! Help is on the way, but these Luxim bulbs are not yet priced for the public. http://www.news.com/1606-2_3-6234653.html?tag=nl.e703

All you opera opposers, (opera not Oprah): Return of the Phantom: It's that phantom electricity that sneaks out of your home from all the gadgets we accumulate. Plug them into a six-port power surge protector/power strip costing approximately **$12.99** and save. With this strip tip it is easy to turn them all off at once with the flip of a switch on the power strip. Even better are the ones supplied with an automatic shut-off switch.

Zip it, button it, tape it, just do it: Remember to throw a blanket around your water heater. If your water heater is old enough that its insulation is fiberglass instead of foam, your bank account will benefit from a water heater blanket from the local hardware or home supply store. To check the insulation, look at the pilot light access in gas models and for electric water heaters, the best access is probably at the thermostat. Cost is about **$18**
http://www.aceee.org/consumerguide/checklist.htm

Use a hose nozzle: Ask a fireman if you don't understand this one. To cut down the amount of water use, purchase a water-spraying nozzle. Free flowing garden hoses can release over 12 gallons a minute.

Top of the line nozzle may cost up to **$18** and will give you more control of the flow and force of water. This is another great way to use water more efficiently. Be sure to seal up any leaky connections and use those little red rubber washers inside every hose connection and at the nozzle. http://e3living.com/water-watch-8-pattern-spray-nozzle

As time rolls by: Is filling tires every four weeks with air too much effort you? Try this tip, for about $5 a tire **($20)** you will be rewarded many times around and around. Nitrogen! According to the Get Nitrogen Institute race cars and airplanes are already filling up on this natural occurring gas. Natural nitrogen is in the air we breathe (21 percent oxygen and 78 percent Nitrogen). By replacing ambient oxygen with nitrogen in your tires, there is less oxidation. Oxidation occurs when oxygen reacts with high temperatures and pressures. It damages the inner lining of the tires, the belt package and even the rim. In fact some tire manufacturers claim that the use of nitrogen will extend the life of the tire by 25 percent. With nitrogen you won't have to fill your tires as often and over the miles you will buy fewer tires.

To see exactly how much money you will save and to find the closest nitrogen supplier in your area visit the Get Nitrogen Institute. So, if you want even better fuel efficiency, longer life on your tires, increased safety by the reduced blowout potential just switch to nitrogen. You can expect to maintain the correct pressure in your tires for almost one full year. Now that is a sweet tip from a saintly friend know as Paul E. Boy.
http://www.getnitrogen.org/sub.php?view=cheaper

Find the biggest sucker with a kilowatt meter: This handy **$25** gadget lets you plug any regular electrical, (no 220), device into the meter. It then measures just how many kilowatt-hours the item is sucking. Once you see where the power is being sucked from the power grid and where your money is going, it is much easier to find a solution.
For instance it was discovered that it is better to brew coffee in the electric pot and turn off the automatic heat source immediately. When you're ready for the

next cup simply heat it in the microwave for one minute. This beats keeping the heating element warm for hours on end and is yet another way to watch your kilowatts drop to less than 500 per month. The challenge has begun. Charge!

G-Wiz: There is a claim that fifty million disposable diapers go into landfills every day. It is estimated each one require 500 years to decompose. If Christopher Columbus' mom had used disposables, they'd be almost decomposed by now. There is a better way and it isn't the old fashion cotton diapers either. The best alternative to the landfill is flushable disposable diapers. G-Diapers have created chlorine free; diapers that, if not flushed, will decompose in 50-150 days in compost. A starter kit costs **$27** and includes two panties and ten earth friendly flushable liners.
http://www.gdiapers.com/happy-planet

Special Occasions: Require special consideration: Wrap that gift with **Eco-Source.** They offer beautiful tree-free wrapping paper. It's made from 40 percent flax, 40 percent hemp, 20 percent recovered cotton and cost **$28.00** per package of 20 sheets; www.islandnet.com/~ecodette/ecosource.htm; 800-665-6944.

Loco-Motion: Don't go crazy there is a solution. Just like the outdoor light sensor, no wiring here either for this easy-to-install indoor motion sensor. Simply unscrew the existing 25W to 100W bulb, screw in the sensor and replace the bulb. The light will stay on for 4 minutes and then turn off automatically. It is incompatible with CFLs but perfect for garage, laundry room or any place where you need light only for a few moments. **$29.99** http://www.smarthome.com/2512.html

Purchase bamboo towels: Dry your dishes and yourself with bamboo towels. They are soft as silk and 70 percent more absorbent than cotton. Bamboo absorbs water faster to dry you faster. It resists microbes and stays hygienically cleaner longer. Towels stay fresh so, less laundry-less energy costs, too. Most are made from 70 percent bamboo and 30 percent cotton. Colors are created with certified low eco-impact dyes. Chlorine and softener-free finishing preserves the purity and natural softness of the bamboo. Towels are available for about **$35** a set. Warning: habit forming and wonderful for planet. The planet, remember, is the home for your house. It's the only one we have. Check out these girls: greengirls3@gmail.com or http://www.thecompanystore.com/parent.asp?product=VF97x&dept%5Fid=3601&cm_ven=NexTag&cm_cat=Towels%20Rugs&cm_pla=BAMBOO_TOWEL_BATH_SH&cm_ite=VF97XLAQU&code=macs=T7NXTG

~GREEN BEINGS –SAVE GENES~

Peel me a grape and fan me with... well uh a paddle ceiling fan: Purchasing fans can be as low as **$35** each. Fans allow you to turn the thermostat up several degrees and still provide a comfortable feeling room. Even using a hand fan helps to move the air and thereby cools your skin. By wearing cooler, lightweight and light color clothing in the summer most people feel comfortable between 72-78 degrees. A ceiling fan extends that parameter upward to 82 degree and you will still feel comfortable. Remember to turn the fan off when you leave the room. A ceiling fan cools you — not the room.

Nobody's favorite -get off the list: Pat on the back to Green Dimes for leading the way toward junk mail freedom at http://www.greendimes.com/index.php You can choose a continuous $4 per month or **$36** dollar per year fee to stop all your junk mail. In addition, this organization will plant a tree on your behalf for each month you are with their service. Now *that* is service with a smile.

Another alternative, **you can pay a small fee of $41 to 41pounds.org**. They promise to completely remove your name from up to 95 percent of the 41 pounds of junk mail you receive. They start by contacting each organization from which you receive mail and/or catalogs. This is a one-time fee. In addition to removing your contact information from the junk mail list, they'll also donate approximately 50 percent of their fee to the New American Dream junk mail campaign! What are you waiting for?

Well connected: With more and more power outages occurring, these handy gadgets will keep you well, connected. Solar charger for IPods or PDA's is available for as little as **$45-$50** at Silicon Solar.
http://www.siliconsolar.com/isol-plus-p-16635.php

Purchase a rain barrel: A rainwater collection system prevents some of the storm-water runoff that reaches and pollutes our rivers. The first inch of rain carries the most pollution from contaminates in the air and land such as animal feces and acid rain. Because of the massive quantity of roof surface, run-off is one of the biggest pollutants. For every 1000 sq. ft. of roof, one down spout can collect 600 gallons of water to use and prevents runoff. This prevents leaky basement, too. The water can serve as a source for your lawn and garden. Most rain barrels hold sixty gallons of water and cost **$45 to $80**. Place a couple under each downspout of your abode. There are many colors and styles from clay urns to plastic ribbed or wooden barrels. Falls Church City Environment Web:

http://www.fallschurchenvironment.org/usingrainbarrels.html Free Directions on making a rain-barrel at Arlington Echo: http://arlingtonecho.net/

Compost Bin: Go organic with an outdoor composter. Make your own rich loam to grow delicious vegetable and a beautiful lawn. Plants thrive in organic matter. These huge bins hold all the food waste from your kitchen compost crock plus you can add leaves and lawn clippings from your yard. There are more than 25 styles of outdoor bins available if you don't like the free version described earlier. Some are round barrels, some square boxes, some resemble a park bench and one design incorporates a flower box on top. Nature does the work by breaking down plant material, rapidly decomposing and magically changing waste to terrific healthy organic matter. Prices range from $39 at Sam's Club to $150 depending on size and style. Rewards take only about two months. Green thumbs on the way.

Chapter Ten

Ulysses

To a Few Franklin's

Microbes with sunglasses? Nope, they are just *cellular shades* for windows. This is the easiest way to increase your R-value. Since most homes lose 25 percent of their heat or cooling through windows, targeting them for improvement is a perfect way to save bucks. Most windows have an R-value that is only somewhere between one and three. But you don't have to install new windows designed by NASA to raise your R-values. Just install cellular shades, which give you R-values from two to nearly five! Cellular shades have air pockets between the layers of fabric that traps air and prevent its escape. They come in single cell, double cell, and triple cell shades. The more cells, the higher the R-value, the greater the savings. It's a great alternative to replacing every window in your home. The price range **is $27 to $45** per window at J.C. Penny.
http://www.energyhawk.com/heat/index.php

No excuses now! Wash your hands with every flush: The latest and greatest in toilet technology is Sink Positive. This is a sink that replaces the lid on your toilet tank allowing you to wash your hands with fresh water before you flush. The extra water drains into the holding tank and is ready for your next flushed. It really isn't that new, it has been used in Asia for many years. All it requires is replacing your current lid on your toilet tank. The cost is about **$90-110** and can be viewed at www.SinkPositive.com or call 615-217-8066

On the go with solar Solio: These handy little gadgets allow you to power all your electronics: mobile phone, MP3 players, GPS, PDA's and so on when you are away from a conventional recharging port. Powered by the everlasting sun, solar chargers can be purchased for **$79 to$199**. This is especially handy for travelers and outdoor enthusiasts. http://www.solio.com/charger

Install a programmable thermostat: This small investment has proven results and big savings. By setting parameter for the thermostat you can save by taking advantage of periods in the day and night when your home doesn't need to be kept at a comfortable temperature. At a cost of **$30-$250** a programmable thermostat can easily save you more than $150 every year. The difference in price is due to the amount of flexibility in programming times. The thirty-dollar version will save you just as well as the more expensive models. Start low and increase only if it will save you additional money.

Ace, King, Queen, Jack, Ten a Full-House power monitor No gambling necessary with this gadget. It will help you save money and reduce more carbon emissions than most other meters. Every home will benefit from its use. For **$135** an easy to install sensor is attached to your outside utility meter. A battery-operated display unit then is placed inside your house where you can view and monitor the kilowatts being used in real time. Some monitors sound an alarm when you exceed a preset amount of electricity. The alarm will buzz until you turn off something to quiet the gadget. This is a great way to cut your electric use. You can watch the charges to your electric bill stop as you turn off the power. What a great way to teach your family just how important each member's role is in the efficiency challenge. Seeing is believing! http://www.smarthomeusa.com/ShopByManufacturer/Blue-Line-Innovations/Item/PCM

BRAINSTORM

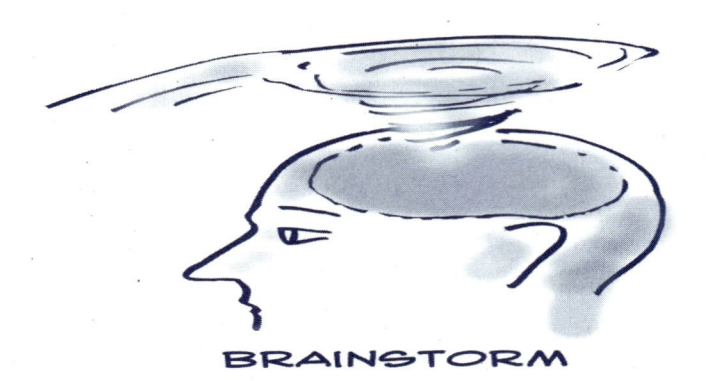

Flat panel computer screens: Replace your old computer monitor with a new flat panel style. The new technology allows computer monitors to take up less space and use a lot less power. It is reported they require only one third the power of conventional monitors. Don't forget to recycle the old monitor. The average cost of a new monitor is approximately **$150**.

Facing South? Any home without an extended overhanging roof can still have passive solar protection with window or door awnings. This alone could save up to 65 percent of cooling costs by installing them. They will block the summer sun's intense heat. Awnings costs $150 to $200 per window or door. Remember to check with salvage companies or second hand construction businesses; often you can pick up the common size frames for a song and then just purchase new canvas. This would cut the costs nearly in half.

The great cover-up: In addition to the global warming issue, another cover up is the solar blanket! This is a novel collection of clear blue bubbles in a blanket. They are similar to the air-filled plastic bubbles that we all love to pop, found in packaging and shipping products. Solar blankets lay on the surface of the pool water. They work by capturing heat from the sun and warming the first few inches of water below the blanket. They have been used for several decades to increase the pool season. If you are using an electric pool heater, a solar blanket is a smart improvement. They also save water by reducing evaporation and minimize nighttime heat loss. Solar blankets are available in many sizes and cost **$90 to $150**.

Better than a flicker in the night: Have you ever notice how a single little candlelight could light up a very dark area? This cool idea does the same. Solar tubes can be installed from your roof to your closet or other similar dark areas of the home like hallways, bathrooms and stairwells. By putting one in all your closets or hallways, you can eliminate the need to turn on the light during the day. There are do-it-yourself kits that claim the process is much easier to install than a

~GREEN BEINGS —SAVE GENES~

standard skylight if you are handy. Most start in the range of **$200.**
http://www.ebuild.com/articles/catCode.-1/articleId.474242.hwx

Captain Cook...yum-yum: Bake it for free. Just purchase a 21-pound completely portable solar oven. Temperatures in the highly polished, mirror-like anodized aluminum can reach 360-400 degrees. There's no chance of fire and the oven can be cleaned quickly and easily with glass cleaner and they will never oxidize or rust. These ovens cost about **$250** and, of course, there is no operating cost. Energy is again free from our sun 365 days per year regardless of the outdoor temperature. Amazing. This is great for summertime bakers and all you campers and boaters, too. To see the guidelines on building one yourself, visit Energy Saving Now web page at: http://sunoven.com/order.asp
http://energy.saving.nu/solarenergy/cookers.shtml

Hogwash? No, white wash: Here is another low budget way to save more. Paint your roof white. We could save $750 million dollars in the US alone if we all had white roofs. White roofs decrease cooling cost simply because white reflects the sun's heat (just like polar ice caps). No, they don't produce electricity but they are energy efficient. It is claimed that white roof temperatures can be 30-90 degrees cooler than conventional black roofs. This will lower individual buildings cooling costs between 20-50 percent. Some argue that dollar-for-dollar white paint is more cost effective than photovoltaic solar panels that cost about one thousand dollars. Because of the continual loss of the white polar ice caps and their replacement of the dark color ocean, the planet is warming faster than first estimated. It is likely by 2013 the ice caps will be totally melted. Some scientist claim it may be necessary for every roof in the world be painted white to reclaim some of the reflective qualities we are losing from the ice shelves and polar caps. For more fascinating thoughts see Mike Tidwell's latest report on this topic at www.orionmagazine.org and Sci/Tech online magazine: http://blogcritics.org/archives/2007/07/02/201753.php

Sweet success: My personal favorite item for really making a difference is using a Nature Mill. At $299 this incredible devise fits into a normal low level kitchen cabinet and pulls out similar to a trashcan. This product offers a practical approach to composting right inside your kitchen or preferred space. This is a great solution to folks unable to compost outdoors. You can put organic food waste and meat and dairy products in this composter. There is no odor, bugs or varmints. After two weeks, presto food waste has turned into rich organic soil, great for all your houseplants, lawn or garden. If you have excess, just sell it in five pound bags as organic soil   Imagine that-selling your garbage? Nature Mill uses five kilowatts per month of energy - or about 50 cents month, less than a garbage truck would burn in diesel fuel to haul the same waste. The mill is made from recycled material. Great invention! http://www.naturemill.com/howItWorks.html

Replacing the A/C window unit; with a new Energy Star model, you will reduce CO_2 emissions by 105 pounds and save about $350. That is based on three months of use or 750 hours.

Attic fans: Invest in an attic fan to keep your home cooler and your roof lasting longer. An attic fan will cut cooling costs by approximately 30 percent because it will rid the house of super hot air that becomes trapped in the attic and backs up into the home. An air conditioner should only be used when the indoor air becomes warmer than 82 degrees. Attic fans will help delay that indoor heat. Attic Fans Dot Com: http://www.atticfans.com **$90 to$400**

Chapter Eleven

As Much As You Choose

Second hand Rose: Buy pre-owned things. Not only is it less expensive, but it also saves raw materials and energy. It's easy to find a thrift store, second hand shop or the best deals of all-can be found at local garage sales. You will be pleasantly surprised by the fantastic bargains. You just need patience and perseverance. Don't forget to check out the national Craig's List mentioned above for other pre-owned deals. http://washingtondc.craigslist.org/. One can also donate wearable business clothing to *Dress For Success*. This organization helps low-income women in their job search. Call 212-532-1922 or visit www.dressforsuccess.org

As mentioned earlier, salvage companies are an excellent source of architectural building supplies. They may have some beauties you just can't find today. Most large cities have demolition or reconstruction supply houses. Many have lists of handymen to hire for installation projects, too.

Brown thumb? No problem, get your food for thought at local farmer markets. Make locally grown foods your first priority while shopping. The average meal in the US travels 1,200 -2000 miles before arriving at your table. Buying locally saves the cost of fuel transporting long distances. Avoid frozen foods, which require ten times more energy to produce and transport. Information on finding local farmer markets or one of the more than 1,200 small US farms that offer fresh produce subscription plans can be found through US Department of Agriculture: www.ams.usda.gov/farmersmarkets/map/htm or visit Union of Concerned Scientists:http://www.ucsUS.org/publications/greentips/page.jsp?itemID=27195279

Go el`natural! Make your own natural cleaning products and stop adding to the toxic waste in our environment. This will save you plenty of money, too. For a perfect all-purpose cleaner, start by making cleaning products with vinegar and baking soda instead of buying hazardous chemicals. This is so inexpensive it is almost free.

Use one teaspoon of liquid soap or (brand name Borax) dissolved in a quart of warm water. For tougher jobs, use one-half cup borax, one-half teaspoon liquid soap, and a splash of vinegar in two gallons of warm water.

Eliminate the need for chemical fabric softeners by adding a quarter cup of vinegar to your washing machine's rinse cycle. Compare the cost of vinegar versus fabric softener the next time you visit the store. Eco Kids: http://www.ecokids.ca/pub/eco_info/topics/waste/itsnotwaste/aboutwaste/whats_in_garbage.cfm
If you prefer to purchase natural non-toxic cleaning agents for your home and your body, check out one of the wellness companies such as Melaleuca for their product line. www.melaleuca.com

Planktos ecosystem restoration: You can help erase your carbon footprint by aiding the restoration of plankton, a vital microscopic plant in the ocean that removes CO_2 from the atmosphere. When CO_2 from smokestacks and tailpipes dissolve into the ocean waters the carbon dissolves and changes form to carbonic acid and lowers the water's ph and threatens the survival of the entire food chain including coral reefs. Today more than 25 percent of all coral reefs are dead or threatened with extinction due to algae that coats the coral and grows so well in acidic water.

Between carbon and phosphates, our most valuable ocean resources are being destroyed. Bet you didn't realize when you you're washing your dishes with phosphate soap, driving your car that emits carbon or turning on your electricity from coal fired power plants you are adding to

the destruction of coral reefs, fish and healthy waterways? We need to become aware first and foremost.
http://www.solarenergylimited.com/content/view/63/40/lang 650-619-0013 david.kubiak@planktos.com

Make your own laundry soap: At three cents per load this is worth the effort: it will save you $70 on 288 loads of laundry. For complete instructions visit the Simple Dollar or try this rendition. You will need:

- 1 bar of soap, pick a smell and color you enjoy
- 1/6 box of Arm and Hammer Washing Soda (Sodium Carbonate)
- 1/4 box of Borax
- 1 five-gallon bucket filled with three gallons of tap water

Bring four cups water to near boil on the stove and add in the shaven bar of soap. Get your five-gallon bucket add the soapy mixture. Stir then add washing soda. Stir while adding the Borax. Mix well. Leave overnight. In the morning just add one cup to your dirty laundry and start washing away.

The Simple Dollar:
http://www.thesimpledollar.com/2007/03/15/how-to-make-your-own-laundry-detergent-and-save-big-money

FYI: Washing soda is stronger than baking soda (Sodium Bicarbonate). It is caustic and alkaline. There are no harmful fumes, but you still need to wear rubber gloves. Adding a little water to the washing soda can make a paste useful for stubborn stains such as petroleum.

Made in the shade: Try using baking soda. It works wonders for scouring everything from your teeth to the kitchen sink. Add a few drops of the laundry detergent above to a few tablespoons of baking soda. This will make a paste that is good for your cleaning your sink.

For your teeth, just use water to make the paste. Vinegar diluted in cool water is great for tile floors, mirrors, coffee pots and even toilets to help remove rust. Add a little of the baking soda again to form a scouring paste.

Murphy's oil soap has a couple of great uses. This is a vegetable-based soap that is good for wood floors. But, did you know if you pour the dirty water on many plants, they are likely to be protected from those leaf sucking scaly bugs. This is wonderful for rose bushes. Bon Ami is another biodegradable scouring agent that will not scratch fine metal surfaces.

~GREEN BEINGS —SAVE GENES~

Make your own pesticides: This is a way to protect the environment, your children and pets. Boric acid has proven to be an effective organic pesticide. Planting garlic chives around your garden will help keep aphids away. Visit the Audubon for a long list of remedies. http://www.audubon.org/bird/at_home/IPM_Alternatives.html

"WELL THERE IS MORE NOXIOUS SMOG AND HAZARDOUS AIR THAT GUARANTEES US MORE ASTHMA AND CANCER PATIENTS IF WE START OUR PRACTICE NEAR, WELL JUST ABOUT ANY SEAPORT IN THE UNITED STATES"

Stop baulking and start caulking: Sealing air leaks are not the only great thing that caulking does. It can prevent water damage inside and outside of the home when applied around faucets, ceiling fixtures, water pipes, drains, bathtubs and other plumbing fixtures. Windows and doors usually require about half a tube or cartridges. Caulking around the foundation or sill of the house will require about four cartridges. Caulking compounds come in other forms such as an aerosol-can; squeeze tubes, and ropes for small jobs or special applications. Here is a handy chart from DOE. So grab your caulking gun and start shooting.

http://www.nrel.gov/docs/fy01osti/28039.pdf

Caulking Compound	Recommended Uses	Comments
Silicone: Household	Seals joints between bath and kitchen fixtures and tile. Forms adhesive for tiles and metal fixtures. Seals metal joints as in plumbing and gutters.	Flexible: cured silicone allows stretch of joints up to three times the normal width or compression to one-half the width.
Silicone: Construction	Seals most dissimilar building materials such as wood and stone, metal flashing, and brick.	Permits joints to stretch or compress. Silicones will stick to painted surfaces, but paint will not adhere to most cured silicones.
Polyurethane, expandable spray foam	Expands when curing: good for larger cracks indoor or outdoors. Use in non-friction areas, as rubber becomes dry and powdery over time.	Spray-foam quickly expands to fit larger, irregular-shaped gaps. Flexible. Can be applied at variable temperatures. Must be painted for exterior use to protect from ultraviolet radiation. Manufacturing process produces greenhouse gases.
Water-based foam sealant	Around window and doorframes in new construction: smaller cracks.	Takes 24 hours to cure. Cures to soft consistency. Water-based foam production does not produce greenhouse gases. Will not over-expand to bend windows (new construction). Must be exposed to air to dry. Not useful for larger gaps, as curing becomes difficult.

Butyl rubber	Seals most dissimilar materials (glass, metal, plastic, wood, and concrete.) Seals around windows and flashing, bonds loose shingles.	Durable 10 or more years: resilient, not brittle. Can be painted after one week curing. Variable shrinkage: may require two applications. Does not adhere well to painted surfaces. Toxic; follow label precautions.
Latex	Seals joints around tub and shower. Fills crack in tile, plaster, glass, and plastic; fills nail holes.	Easy to use. Seams can be trimmed or smoothed with moist finger or tool. Water resistant when dry. Can be sanded and painted. Less elastic than above materials. Varied durability, 2-10 years. Will not adhere to metal. Little flexibility once cured. Needs to be painted when used on exteriors.
Oil or resin-based	Seals exterior seams and joints on building materials.	Readily available. Least expensive of the four types. Rope and tube form available. Oils dry out and cause material to harden and fall out. Low durability, 1-4 years. Poor adhesion to porous surfaces like masonry. Should be painted. Can be toxic (check label). Limited temperature range.

Stripping for hot weather! Oops, weather-stripping for cold weather. Hot or cold, proper weather stripping around your windows and doors will save ten to twenty percent off your heating and air conditioning bills. Though windows and doors are an obvious source of air exchanges, the worst problems are usually gaps around utility cut throughs called plumbing penetrations. Don't forget the gaps around the chimneys and all the

electrical outlets in each outer wall of the house. Remove the faceplate around electric plug outlets and switches. Then fill the gap with an insulating material.
Did you know cobwebs are evidence of moving air? If you see them in your basement you can be sure there is air movement. Find the air leak and seal it before cold weather. Edison Electric Institute:
http://www.eei.org/industry_issues/retail_services_and_delivery/wise_energy_use/100Ways.pdf

Apply weather stripping to a clean dry surface greater than 20 degrees. To determining the length of weather stripping you'll need, measure the perimeter of the windows and doors. Then add approximately ten percent. Because weather stripping comes in so many sizes, widths and consistency, the following chart is included in this text with thanks to the US Dept. of Energy. The products are chosen carefully for location and ability to withstand friction and temperature changes.
http://www.eere.energy.gov/consumer/your_home/insulation_airsealing/index.cfm/mytopic=11280

Weather stripping	Best Uses	Comments
Tension seal Self-stick plastic (vinyl):	Inside the track of a double-hung or sliding window, top and sides of door.	Surfaces must be flat and smooth for vinyl. Can be difficult to install, as corners must be snug. Consult website for more detailed explanation.
Felt: Plain or reinforced with a flexible metal strip; sold in rolls. Must be stapled, glued, or tacked into place. Seals best if staples are parallel to length of the strip.	Around a door or window (reinforced felt), fitted into a doorjamb so the door presses against it.	Low durability; least effective preventing airflow. Do not use where exposed to moisture or where there is friction or abrasion. All wool felt is more durable and more expensive. Very visible.
Reinforced foam Closed-cell foam attached to wood or metal strips.	Door or window stops; bottom or top of window sash; bottom of door.	Can be difficult to install, must be sawed, nailed, and painted. Very visible. Manufacturing process produces greenhouse gas emissions.
Tape: Nonporous, closed-cell foam, open-cell foam, or EDPM (Ethylene Propylene Diene Monomer) rubber.	Top and bottom of window sash; doorframes; attic hatches and inoperable windows. Good for blocking corners and irregular cracks.	Durability varies with material used, but not especially high for all; use where little wear is expected; visible.
Rolled or reinforced vinyl Pliable or rigid strip gasket (attached to wood or metal strips.)	Door or window stops; top or bottom of window sash; bottom of a door (rigid strip only).	Visible.
Door sweep: Aluminum or stainless steel with brush of plastic, vinyl, sponge, or felt.	Bottom of interior side of in-swinging door; bottom of exterior side of exterior-swinging door.	Visible. Can drag on carpet. Automatic sweeps are more expensive and can require a small pause once door is unlatched before retracting.

Magnetic: Works similarly to refrigerator gaskets.	Top and sides of doors; double-hung and sliding window channels.	
Tubular rubber and vinyl: Door or window presses against them to form a seal.	Around a door.	Self-stick versions challenging to install.
Reinforced silicone: Tubular gasket attached to a metal strip that resembles reinforced tubular vinyl	On a doorjamb or a window stop.	Installation can be tricky. Hacksaw required to cut metal; butting corners pose a challenge.
Door shoe: to protect under the door.	To seal space beneath door.	Fairly expensive, installation moderately difficult. Door bottom planning possibly required.
Bulb threshold: Vinyl and aluminum	Door thresholds	Wears from foot traffic, relatively expensive.
"Frost-brake" threshold	To seal beneath a door.	Moderately difficult to install, involves threshold replacement.
Fin seal: Pile weather-strip with plastic Mylar fin centered in pile.	For aluminum sliding windows and sliding glass doors.	Can be difficult to install.
Interlocking metal channels: Enables sash to engage one another when closed	Around door perimeters.	Very difficult to install as alignment is critical. To be installed by a professional only.

~GREEN BEINGS —SAVE GENES~

Save the rainforest, save a brother, save yourself: Only two percent of the Earth's surface remains a rainforest ecosystem, yet more than 50 percent of all plant and animal species live there. By purchasing only rainforest protected Fair Trade labeled products such as coffee, tea, chocolate, bananas, mangoes, pineapples you are indirectly improving your own air quality. You are preventing more erosion and water pollution. Saving the rainforest helps keep the atmosphere and the ocean waters cooler thereby reducing the number and intensity of hurricanes and storms.

The rainforest protects wildlife and fish, and holds great potential for the discovery of lifesaving medications among its diversity. These undiscovered medications and treatments could actually save *your* own life or that of a loved one in the future.

In addition, when you buy products carrying the Fair Trade certified label products you are helping to empower families in developing communities. This certification ensures the grower will receive a fair price for their labor as well as protecting the environment. Fair Trade also keeps the health, home, and education of the farmer and his family as a top priority. Fair Trade farmers can be to be successful and productive in their own country. No need to immigrate to find work to

support a family. Last but not least, you are eating the healthiest food produced on the planet. It is straight from their garden to your home. No bugs attached. It is a win-win situation. Fair Trade Federation:
http://www.fairtradefederation.org/
http://www.coopamerica.org/programs/fairtrade/products/index.cfm

So on your next trip to your grocery store, why not stop by the store manager and let him know that you want Fair Trade products available. Yes, it currently cost a little more, but the law of supply and demand will bring the price down as more and more people ask for it.
Oh, you may wonder what the price is for a pound of Fair Trade coffee. Coffee farmers trading in the Fair Trade system receive a floor price of $1.31 per pound, or $1.51 per pound for duel certified Fair Trade Certified™ and Organic certification. This is nearly double what conventional farmers currently receive. Please support Fair Trade products when you shop.

While sipping your delicious Fair Trade coffee or tea, please do so from a Disappearing Continents Mug. Take this with you where ever you go. This mug when filled with hot liquid suddenly transforms the current map of the world to reveal the predicted shorelines as glaciers melt and sea levels rise. This leads right back to chapter four in the discussion of home gardening.

Organic Consumers Organization: You will find plenty of examples for protecting your own body as well as protecting the planet. The scientific community has demonstrated in multiple studies that pesticides and other toxins in foods adversely affect pregnant women, children, the elderly and probably all those in between. It is alarming when one becomes aware of the contents of foods we have trusted to be healthy.

For instance the 'dirty dozen' are most shocking. Environmental Working Group (EWG) based results of nearly 43,000 tests for pesticides on produce collected by the US Department of Agriculture and the US Food and Drug Administration between 2000 and 2005. They have reported 43 fresh fruits and vegetables contaminated with pesticides. Listed here are the biggest culprits.

Since 2005 there have been numerous episodes of contaminated foods resulting in the death of many unsuspecting people. Lettuce and strawberries and in the US this summer the tomato and jalapeño pepper are on the do not eat list. http://www.organicconsumers.org

Visit *Food News* for detailed information on pesticides and foods. What is good for you is good for the planet too. Avoid purchasing cut flowers, houseplants or

landscape products that are laden with pesticides. They might be pretty for a short time. But the earth will be prettier without their use for a long time.
http://www.foodnews.org/reduce.php

Dirty Dozen		
FRUIT OR VEGGIE	RANK (1=worst)	Pesticide Load (100= highest)
Peaches	1	100
Apples	2	96
Sweet Bell Peppers	3	86
Celery	4	85
Nectarines	5	84
Strawberries	6	83
Cherries	7	75
Lettuce	8	69
Grapes - Imported	9	68
Pears	10	65
Spinach	11	60
Potatoes	12	58

Young McDonald had a farm, C A C A F O: Oh, let me bring you up to date. It is a careless, cruel, inhumane, environmentally horrific farming method that is called Confined Animal Feeding Organization. What does Bessie have in common with global warming? Besides the facts that Americans love their beef as much as their fuel consuming cars and are just as resistant to giving up either? This section is here to inform you of the environmental cost of eating meat and solutions besides the soy burger.

Through no fault of her own: Yes, dear Bessie, the brown-eyed cow, surprisingly may be the cause of more environmental havoc than even burning fossil fuel. The facts are there for anyone who chooses to look at the big picture. Sorry to burst another bubble, but our consumption of meat is just as hazardous as oil. You may say to yourself, "Prove it and tell me why".
Don't worry, I am not advocating that we all become vegetarians, but there is a wiser and healthier way to produce and consume meats and dairy products.

First here are some the positive and negative facts:

- ◆ (+) A confined feeding operation is a method of housing animals in one site to produce as much meat or dairy at the lowest possible cost. Yeah that is a positive + for Efficiency.

- ◆ (+) Can you just imagine the quantity of the possible biofuel potential?

- ◆ (-) One dairy cow annually produces more gas pollutants than one car or light trucks. When a cow belches or 'passes gas' from intestines the gas released is methane, a green house gas that is 20 percent more harmful than carbon.

- ◆ (-) Eating just 2.2 pounds of beef is equivalent to driving a car for 155 miles. The energy required to bring those two pounds of meat to you is also equivalent to burning the old 100- watt incandescent light bulb for 20 days. Uh oh, we may be in trouble with our taste buds!

- ◆ (-) Confined crowded living conditions for people or livestock are a natural incubator for disease.

- ◆ (-) Pesticides are used in and around the animals to keep diseases at bay. I wonder if cows get asthma.

- ◆ (-) Animals are all fed antibiotics to prevent illness. This is believed to be one of the main causes for human insensitivity to antibiotics. When we ingest meat from animals that are full of antibiotics, that cellular memory is transferred to our own human immune system.

- ◆ (-) Cows also cannot digest corn, which is what they are fed in feedlots. Corn fed cow meat is high in Omega 6 not the desirable Omega 3.

- ◆ (-) Pasture raised grass-fed cows are an Omega 3 fat source when eaten.

- ◆ (-) The animals are not allowed to graze or go to pasture to eat natural vegetation. What they have been fed in the past is gruesome and has been called 'rendering' so not to sound so, well shocking. However, other deceased animals have been ground up and placed back in the cows food supply. Wait I know cows are suppose to be vegetarians by choice. But confined they have no choice.

Well, yes it is gruesome but it gets worse. Rendering can lead to diseases such as the famous Mad Cow disease. I'd be mad too if I were confined, injected with hormones, antibiotics, steroids, and fed the remains of my dead relatives. "Scientists aren't exactly sure how cows get this disease, but it probably happens when cows eat feed made from cows infected with mad cow disease" according to the Food and Drug Administration. http://www.fda.gov/oc/opacom/kids/html/madcow.htm Some ranchers cut costs by feeding livestock ground-up dead animals, animals that have died of disease - as well as humane society animals, cats and dogs that have been euthanized. Also, being used as feed were road-kill and parts of animals not fit for human consumption. This process of grinding up diseased, dead animals for feed for other animals is somewhat controversial according to Dr. Lorraine Day. http://www.drday.com/madcow.htm

- ◆ (-) Only 40 of the 200 counties around the world that have CFOs are capable of responding to disease such as Mad Cow Disease, Avian Flu, West Nile Virus, Bluetongue, and Foot and Mouth disease.

- ◆ (-) In addition to the multiple antibiotics, cows for instance also receive injections of hormones and steroids to increase size and weight.

- ◆ (-) Confined animals are equal to confined waste products. A CFO usually confines 1000 or more animals in one spot.

- ◆ (-) Unfortunately most waste finds its way into the waterways that cause tremendous fish death and water pollution.

- ◆ (-) The world's total meat supply was 71 million tons in 1961. In 2007, it was estimated to be 284 million tons.

- ◆ (-) Nearly 10 billion animals are confined and slaughtered every year.

- ◆ (-) More meat means a corresponding increase in demand for feed, especially corn and soy. Now think of fixing dinner for ten billion animals!

- ◆ (-) Americans represent five percent of the world population and eat about eight ounces a day or 15 percent of the world's total meats. That is twice the global average.

Solutions: If Americans were to reduce meat consumption by just 20 percent it would be as if we all switched from a gas-guzzling car to an ultra-efficient auto such as the Prius.

If every American had one meat-free meal per week, it would be the same as taking more than five million cars off our roads.

Eat Healthy Meat: The American Grass Fed Association (AGA) standards are based on four precepts: total forage diet, no confinement, no antibiotics and no added hormones. American Grass-fed http://www.americangrassfed.org/ And there is an added benefit that milk, meat and eggs from grass-fed

animals are both lower in saturated fats and contain higher levels of heart-healthy and cancer fighting omega-3 fatty acids. Eat Wild web site includes dozens of resources and facts for eating healthy food.
http://www.eatwild.com/resources.html

Worried about not getting enough protein? It is important to correct this misconception. Many people are afraid that following a pure vegetarian diet will not supply all the protein they need. The misinformation of "incomplete proteins" from plant sources is just not true. A vegetarian diet based around any single one, or combination, of these unprocessed starches (rice, corn, potatoes, beans, etc.) with the addition of vegetables and fruits supplies all the protein, amino acids, essential fats, minerals, and vitamins (with the exception of vitamin B12) necessary for excellent health. One of the oldest recipes called succotash contains corn and beans that supply all the essential amino acids. Amino acids are the building blocks of protein. http://www.all-creatures.org/health/plantfoods.html

Remember, "pay now or pay later". Paying the price for healthy food is much more enjoyable than paying the price for prescription medication, radiation and chemotherapy from eating unhealthy foods. This site takes you directly to the state-by-state locator for healthy food that is much less expensive than the health care required treating disease.
http://www.eatwild.com/products/index.html#states

http://www.nytimes.com/2008/01/27/weekinreview/27bittman.html New York Time, *The Meat Guzzler*
Be aware of these facts in order to make the smartest decision; Environmental Defense Fund has extensively studied the carbon facts of local or shipped and trucked foods. For example:

◆ Trucking tomatoes from Spain during the winter produces less greenhouse gas emissions than growing them in heated greenhouses in Britain.

◆ Lamb raised on New Zealand's lush pastures and shipped 11,000 miles by boat to Britain produced 1,520 pounds of CO_2 per ton, while British lamb produced 6,280 pounds. The reason? British pastures provide poorer grazing, forcing farmers to use feed.

http://environmentaldefenseblogs.org/climate411/2007/10/11/food_miles
http://www.edf.org/page.cfm?tagID=20927
http://www.fightglobalwarming.com/page.cfm?tagID=1302

Windows and doors: Certified by Energy Star bases its grading system on U-factor and Solar Heat Gain Coefficient (SHGC) rating (see glossary). When the window is selected for the direction of the exposure, one can minimize heating, cooling and lighting costs. It is recommended that windows on the east-west and north-facing walls have a high SHGC >0.6 to collect solar heat gain in the cold climates. Skylights should be utilized for adequate day lighting. In warm climates, east- and west-facing windows should have a low SHGC and/or be shaded. North-facing windows in warm climates collect little solar heat, so they are used just to provide natural lighting.

When upgrading your windows, look for windows approved by the NFRC (National Fenestration Rating Council). This is a nonprofit voluntary program that tests, certifies and labels windows, doors and skylights based on their energy performance. This label offers the consumer a way to compare products.

Another unbiased opinion is offered through Efficient Windows Collaborative: http://www.efficientwindows.org
The Alliance to Save Energy has an excellent web site allowing one to obtain more information.
http://earth911.org/blog/2007/04/02/energy-conservation-tips-for-your-windows

Computer update: Consider a laptop. These use only about one fifth as much power as a desktop PC. Sierra Club green.life@sierraclub.org

Storm doors: Adding a storm door can be a good investment if your existing door is old but still in good condition. However, adding a storm door to a newer, insulated door is not generally worth the expense since you won't save much more energy. According to the US Department of Energy, the R-values of most steel and fiberglass-clad entry doors range from R-5 to R-6. http://www.eere.energy.gov/consumer/your_home/insulation_airsealing/index.cfm/mytopic=11340

Lights, camera action: All you do-it-yourself film producers can apply film to your windows and watch your savings go up. Though most glazing on windows and tinting are completed at the manufacturing plant, there are kits that allow you to retro fit this stuff.

Adding tinted film to your glass helps to absorb a large amount of the incoming solar heat gain. The film is said to last ten to fifteen years without chipping or peeling. This is inexpensive compared to replacing the entire window.

Blue and green-tinted windows offer greater penetration of visible light and slightly reduced heat transfer compared with other colors of tinted glass. In hot climates, black-tinted glass should be avoided because it absorbs more light than heat. Be wary of reflective window glazing-it absorbs more visible light than heat. This makes the room much darker but reduces the solar gain and lower cooling cost maybe offset by the need for additional electric lighting. US Department of Energy: http://www.eere.energy.gov/consumer/your_home/windows_doors_skylights/index.cfm/mytopic=13410

Roll up the carpet and replace with bamboo flooring: This is the most environmentally friendly floor you can put in your home. Bamboo is the fastest growing grass on the planet. It reaches maturity in just about five years, compared to a hardwood tree reaching maturity in 50-100 years. Bamboo can be manufactured in many colors and shades to resemble hardwood flooring. Bamboo cost $2 to $4 a linear foot compared to hardwood that can easily cost twice as much. This is yet another way to protect our environment.

Just like bath water: That will be your swimming pool when you use solar heat. Go Solar Company

in Southern California claims a typical pool season will cost in excess of $2,000 of natural gas to heat the pool. Installing a solar swimming pool heating system with a life expectancy of over 20 years at an average cost of $4,800 makes economic sense, too. Solar pool heating installation will save a substantial amount of money while providing a warm and comfortable pool. Don't forget to top it with a solar blanket previously discussed.
http://www.solarexpert.com/poolheat.html

Insulation: The easiest and most cost-effective way to insulate your home is to add insulation in the attic. If you have less than six or seven inches, you will probably benefit by adding more. Most US homes should have between R-38 and R-49 attic insulation. In order to achieve this, many homeowners should add between R-19 to R-30 insulation (about 6 to 10 inches). There are many types of insulation for many places in your home. Visit the web page of the US Dept. of Energy for a detailed description chart when shopping for insulation in your home.

According to the Department of Energy, and Energy Star products, if just one in ten homes used energy-efficient appliances, it would be equivalent to planting 1.7 million new acres of trees. Energy Star-qualified appliances use ten to fifty percent less energy and water than standard models. Consumer Report offers Greener Choices at

~GREEN BEINGS —SAVE GENES~

http://www.greenerchoices.org/globalwarmingathome.cfm

Chapter Twelve

Premeditated Assault on Carbon
William and Beyond

William McKinley

Win~win: For every $1 invested in energy efficiency $4 is recognized in savings. And not only do we save money when we invest in efficiency, but we also reduce environmental hazards making our home more comfortable and valuable, reduce overall energy costs, improve the reliability of the utility system, delay the need to build new power plants, slow rising energy prices, create new jobs and strengthen the economy for society as a whole. What more reasons do you need?

Don't be thankless...go tankless! When replacing you water heater, think of installing a tankless on-demand system. These small wonders are around 30 percent

more efficient than a conventional water heater and lasts longer too! The average carbon savings is 3,285 pounds of carbon from the environment each year. If you choose a tankless unit, look for one eligible for federal tax credits EF 0.80 (gas) or EF 2 (electric) because older models may have difficulty responding to multiple demands at the same time, especially in the colder months. Merchant Circle Sims Plumbing Company: Costs varies from **$200-$600**
http://www.merchantcircle.com/blogs/Sims.Plumbing.Company.inc.515-262-6400
http://www.merchanticircle.com/blogs/Sis.Publishing.Company.inc 515-262-6400

Solar attic fan: The best way to decrease the work of your air conditioner in the summer is to ventilate your attic. Temperatures can exceed 160 degrees in the summer. An attic fan can remove 50 degrees easily. By using a solar free power fan you are further helping the environment. A ten-watt fan can ventilate 1200 square foot attic and costs approximately **$430.**

Replace china toilet bowls: Don't forget to recycle the old one when you replace it. They are crushed and incorporated into new road construction or for repairing existing roadways. Replacing toilets as new as 1993 will save the average family over $1,500 per year. When replacing the toilet strive for 1-1.5 gallon

tank capacity. If your toilet predates 1980, replacement is recommended. You can save $2,837 for a family of four over the course of ten years. Initial costs are about **$800**. There are still more ways to improve efficiency with high-pressurized water flushing toilets or even natural composting toilets. Visit Seattle-Utilities: 206 648 7283 or
http://www.savingwater.org/docs/ToiletBrochure.pdf

Whole house fan aka swamp cooler: These are especially useful in hot dry climates. A whole house fan only uses about 25 percent of the power of a central air-conditioning system. A whole house fan is most effective when outside air temperatures are about 82°F. Some caution must be used when operating a whole house fan. It doesn't take real brain power on this one...turn off heating and air-conditioning, open windows, no fires in fireplace and then turn on the whole house fan, **$1,100 to $1,200**.

Solar water heaters: Many claim these to be the best use of solar energy. They use energy from the sun to heat water. The initial cost of a solar water heater is higher than a gas water heater, but the operating heater costs are easy to recognize which is the best value. Solar water heaters definitely save lots of cash over the long term. They are reliable and compete very well with electric and propane water heaters on a cost basis. They obviously prevent carbon pollution. Costs start just over **$1,200** http://www.daviddarling.info/encyclopedia/S/AE_solar_water_heater_cost.html

Replace central air-conditioning: Units with a SEER rating of ten or less should be replaced. A SEER rating of 13 or more will definitely save 25 percent on your cooling bill. By choosing a new Energy Star qualified unit with a SEER of 14, you can reduce CO_2 emissions by about 1,540 pounds annually, assuming it's on for 1,320 cooling hours, or eight hours a day for five or six months. And make sure that your contractor does the sizing calculations so you don't install a unit that's too big-it will cost more and require larger ducts to handle its higher airflow. Edison Electric Institute: http://www.eei.org/industry_issues/retail_services_and_delivery/wise_energy_use/100Ways.pdf

HMMM I CAN ESCAPE THIS NONSENCE BY INVESTING IN SMART ENERGY FOR MY HOME. SUDDENLY MY HOME IS WORTH MORE AND ME TOO.

Replacing your old gas furnace with a new Geo-Exchange unit will save you between 30-50 percent on heating and cooling bills. When buying a new furnace or boiler, make sure you purchase one with a more efficient Adjusted Fuel Utilization Efficiency (AFUE) (see glossary). Starting price is about **$1,400** Energy Star Dot Gov. Energy Star approved furnaces have AFUE ratings of 90 percent or more. Edison Electric Institute: http://www.eei.org/industry_issues/retail_services_and_delivery/wise_energy_use/100Ways.pdf

~GREEN BEINGS —SAVE GENES~

Replace clothes washer: By replacing a top-loading washer with a front-loader energy efficient washer, you will save 30 percent on electric bills. If you choose an Energy Star over a conventional model you will reduce carbon emissions by 356 pounds based on 392 loads. Energy Star washers use only 18-25 gallons of water per full load in comparison to the 40 gallons of conventional washers. Check out Consumer Report on green ratings for washers. An Energy Star unit can cost **$1,500** and uses about 294 kWh compared to an old washer that uses 1051 kWh.
http://www.consumersearch.com/www/house_and_home/washing-machine-reviews/index.html
http://www.powerisinyourhands.org

Even greener than that: Bring that crescent moon back in the house! Natural Composting Toilets are completely odor free, energy efficient and quite comfortable. ☺Most options are waterless, low flow 1-pint-per-flush or vacuum. Visit Envirolet Composting Toilets by Sancor. http://compostguide.com/info are a little costly, **$1,600** but, heck, you will never have to call Roto-Rooter or use nasty drain opening products. They are perfect when you don't want to add a bigger septic tank to your existing property.

Replacing a refrigerator: If your frig is 18 years or older, replace it because it uses 857 kilowatts per hour and a new one will save you 35 percent on your electric bill. If you purchase a high-efficiency model like Energy Star that uses 512 kWh you will reduce CO_2 in the atmosphere by 100-300 pounds. The refrigerator is the third largest energy user in the home. Top of the line Jenn-Air 22 cubic foot costs **$2,200**

Cook tops: Induction elements are the newest and most innovative type of cook top to come along. The induction elements transfer electromagnetic energy directly to the pan where the heat is needed. They are very energy efficient, using less than half as much energy as standard electric coil elements. One catch is that they work only with ferrous metal cookware (cast iron, stainless steel, enameled iron). When the pan is removed, there is almost no lingering heat on the cook top. Currently, induction elements are available only with the highest-priced cook tops. A 36 inch five burner cook top costs a whopping **$2,999** with free shipping. ("Oh Boy", you say. I hear you even through polluted air space). American Council for Energy Efficient Economy: http://www.aceee.org/consumerguide/cooking.htm

Countertops: Recycled glass is being used to produce beautiful countertops containing as high as 85 percent recycled product. There is some competition out there so be sure to check around. California seems to have the lead again on this product. Costs vary but one estimate obtained was $27.50/sq ft and up. There are many uses for recycled glass. Check out this site for a location near you and read about complete comparison analysis of glass to marble or other familiar products. Don't purchase counter tops containing formaldehyde or other gas producing chemicals that costs $85 to $150 per sq ft installed.
http://www.amicusgreen.com/v1/kitchenandbath10tips.htm http://www.vetrazzo.com/products_intro.html

Many companies are now offering environmental building supplies. Some flooring products include:

Linoleum: True linoleum is not vinyl, rather it is a natural material made of linseed oil, pine resin, wood or cork floor, limestone and pigments. Linoleum is coined "the 40-year floor" and is naturally anti-microbial, highly stain resistant and easy to maintain, costs $4 sq ft. Vinyl costs $2 sq ft but last only 15 years and contributes to dangerous organic compounds in the air and in your home. Is it really a mystery that there is such an increase in asthma over the past fifty years? It is the one disease that progresses in spite of medicine.

Cork: Harvested for centuries from the 300-year old living Cork Oak tree (sound familiar?). Yes, it is the same cork that has been used in wine stoppers. Its' soft, cushiony feel is perfect for floor tiles and has been used for insulation panels on space shuttles.

Kirei: This very green building product made from sorghum stalks is a great substitute when building furniture or cabinetry and used frequently for flooring. They grow quickly and are a 100 percent reclaimed agricultural by-product. The boards are strong, lightweight, durable and very environmentally friendly.

Terrazzo: Enviro-glass brings the ancient craft of Terrazzo into the 21st century by combining lightweight epoxy resin with multi-colored glass chips from discarded bottles, mirrors and plate windows or chips from discarded toilets, sinks and tubs to achieve a surface harder than traditional marble Terrazzo. According to Enviro-glass this is the only finish product for less than $1 sq. ft. Good job! http://www.enviroglasproducts.com/terrazzo.asp

Bamboo: Timber bamboo grows mainly in China and Vietnam and can be harvested every four years without destroying the root system. Strips from the stalk's perimeter are laminated into a flat or vertical grain floor plank. This attractive flooring is harder and more stable than some wood.

Some bamboo species could be grown and thrive in the US but producing bamboo timber is labor-intensive. It is thought that paying for transportation from China is more cost-effective than paying for US labor costs. That could be debatable with rising fuel costs and higher unemployment rates in the US.

However, when considering the CO_2 footprint it is usually smarter to purchase lumber from local cooperatives that harvest trees in an ecological and sustainable manner than purchasing bamboo flooring.

Another consideration concerns the amount of formaldehyde-based adhesive used to laminate the bamboo. LEED stand requires no more than .05 parts per million (ppm) of formaldehyde is allowed according to Greenguard Environmental Institute, an independent nonprofit group that certifies the indoor quality of products. http://www.greenguard.org/

Certified wood flooring: The majority of hardwood floors, labeled Forest Stewardship Council (FSC)-certified, come from a third-party certified, well-managed forest. These forests are managed to look and function naturally. They are monitored for wildlife habitat and biodiversity, while supporting local community.

Great old reclaimed wood: Reclaimed wood has real character and often-higher quality than what you buy in home stores today. Most of this flooring is oak, chestnut, Eastern white pine, heart pine, hemlock and other exceptional quality woods. Wood is taken from old barns; antique structures, rivers and lakes, railroad trestles and much of the wood had originated from old growth forest. http://www.trestlewood.com Sometimes reclaimed wood fibers or rice hulls are combined with recycled resins to form composite boards that take screws and nails just like wood but last longer and don't require toxic stains.

Carpeting: While looking for environmentally friendly carpet, be sure to search for Sisal, Coir and Sea-Grass. Wool and recycled fibers are another source of sustainable resources and often utilized.

Real costs of flying with the iron bird: Jet-fueled flights are one of the fastest growing sources of greenhouse gas emissions. Jet fuel uses kerosene, because it packs the most energy per pound. It takes almost ten gallons of fossil fuel to make one gallon of kerosene jet fuel. And to make matters worse, the pollution is deposited in the worst place where no possible carbon absorption can take place, namely high up in the atmosphere. Placing that amount of carbon high in the atmosphere is analogous to placing "salt in the wound". Ouch! Cutting down on long-distance travel would be helpful. Try using a train, bus or bike for at least one of your trips.

Green Flights have just been introduced and are using 'frying fuel' for their experimental biofuel jet. The first experimental flights were made from Reno, Nevada, to Orlando, Florida. Eventually, they hope to fly a similarly fuelled plane around the world.
http://www.sciam.com/article.cfm?id=biodiesel-takes-to-the-sky

When considering travel and the price of fuel, frequently it is more cost effective and carbon saving for one person traveling alone to fly on the standard commercial aircraft than it is to drive alone. Those benefits are negated when two or more people are traveling together, then driving still remains more cost-effective especially in a Prius.

Job well done! You are saving money and securing a home for your children and grandchildren. Just take another look at your initial electric bill. Fill in the blanks below and then compare to your previous bills to see what you have done single handedly! You are a hero! You are more than two-thirds of the way to becoming totally environmentally conscious. You have green in your pocket and your friends are green with envy. Keep on going green!

- Month:
- Utility:
- Costs:
- Kilowatts Used:

PHASE THREE

Chapter Thirteen

Empowering Your Home

Energy: In the US, energy is currently generated from five major sources: coal, nuclear, solar, wind, and geothermal sources. Let's take a closer look at a few of our options.

Coal: Burning coal is the major source of carbon pollution into the atmosphere. Coal was formed 4,000-100,000 years ago from the decomposition of ancient plants and animals. As they decomposed, a sedimentary organic rock was developed and as it aged it became blacker and harder. Coal is classified according to its age and carbon content. Coal contains between 40-90 percent carbons. The youngest is known as peat and is commonly used as fuel in Ireland. The oldest is known as bituminous. This happens to be a common name for roadways in Australia. Coal is found in 38 states in the US; Wyoming, Kentucky and West Virginia remain the top three producing about 59 percent of the coal in the US.

HE KEEPS GROWING

Just One 500 megawatt coal plant releases into the atmosphere each year:

- 10,000 tons of sulfur dioxide (SO_2)

- 10,200 tons of nitrogen oxide (NO_2)

- 3 million tons of carbon dioxide (CO_2)

- 500 tons of small particles such as soot

- 220 tons of hydrocarbons

- 720 tons of carbon monoxide

- 125.000 tons of ash

~GREEN BEINGS —SAVE GENES~

- 193,000 tons of sludge

- 225 pounds of arsenic

- 114 pounds of lead

- 4 pounds of Cadmium

- Trace elements of Mercury and Uranium.

More than 70 percent of the heat generated by coal comes from carbon. Heat naturally rises. Heat and carbon become trapped in the atmosphere and pollution continually rises. Union of Concerned Scientists; How Coal Works:
http://www.ucsusa.org/clean_energy/fossil_fuels/offmen-how-coal-works.html

Most people have heard of 'coal miner's lung'. You know the black lung that is responsible for the deadly and devastating cancer destroying the lives of so many hard working coal miners? Here's another human health analogy to ponder: the atmosphere surrounding the earth acts like the lung of the planet. Just as coal dust or smoke from tobacco can penetrate the human lung after years of exposure and often without evidence of harm until a point of no return. We are seeing the same effect on a much larger scale and a real life form on the planet.

There is a tipping point. For those skeptics who claim, "people have the audacity to think that we could really be responsible for the changes in the atmosphere and actually have caused the rising temperatures on the planet", please think a little deeper. There is a reaction for every action. There is an effect of every cause. Scientists across the globe accept the evidence. There is a limit to parts per million of pollutants our atmosphere can hold without affecting life sooner or later. Coal is one of the dirtiest and devastating assaults on humans and Mother Nature. Just like tobacco it is a very bad and addictive habit. "It's not nice to rule (or fool) Mother Nature". She always wins.

Not only do the top three pollutants come from coal, they also come from nuclear plants. Nuclear is slightly less toxic (if there is such a thing). For example, each megawatt of coal-generated power will be compared head-to-head with nuclear power. Here are some comparisons. Nuclear may appear a smart move but hold your opinion for a few more pages.

> NO_2=1.23 pounds for **coal** electricity vs. 0.77 pounds for **nuclear** energy.
> SO_2= 4.5 pounds for **coal** generated electricity vs. 2.8 pounds for **nuclear** energy.
> CO_2 = 710 pounds for **coal** electricity vs. 443 pounds for **nuclear** energy.

~GREEN BEINGS —SAVE GENES~

For each of these pollutants, EPA has established national air quality standards to protect public health. Ground-level ozone and airborne particles are the two pollutants that pose the greatest threat to human health in this country.
http://airnow.gov/index.cfm?action=static.aqi

Let's talk about air quality and AQI. What? Air quality index is what you should know. The AQI is an index from zero to 500 for reporting daily air quality. It tells you how clean or polluted your air is, and what associated health effects might be a concern for you. EPA calculates the AQI for five major air pollutants regulated by the Clean Air Act: ground-level ozone, particle pollution (also known as PM particulate matter), carbon monoxide, sulfur dioxide, and nitrogen dioxide.

An AQI of 100 for $PM_{2.5}$ corresponds to a level of 40 micrograms per cubic meter (averaged over 24 hours). The $PM_{2.5}$ is a particle size one third that of human hair. The main sources are from emissions of vehicle exhaust, agricultural activity and of course, coal fired power plants. Over 2,000 studies since 1997 link fine particle pollution to strokes, heart disease, respiratory ailments, and premature death. Union of Concerned Scientists among others claims the EPA new air pollution standards do not sufficiently protect the public.
http://www.ucsusa.org/scientific_integrity/interference/epa-particulate-matter.html

Air Quality Index

When the AQI is in this range:	...Air quality conditions are:	...As symbolized by this color:
0 to 50	Good	Green
51 to 100	Moderate	Yellow
101 to 150	Unhealthy for Sensitive Groups	Orange
151 to 200	Unhealthy	Red
201 to 300	Very Unhealthy	Purple
301 to 500	Hazardous	Maroon

Nuclear power: Nuclear power may give the impression of being a wiser choice. In the medical world, health care professionals laugh at reports generated from the nuclear medicine department. We call it "un-clear medicine". Just reversing the first two letters gives a better idea of what we are up against in the energy world, too. Let's take a close look.

Nuclear energy is the most expensive and the least safe of all the non-renewable energy sources. Each nuclear plant will generate twenty tons of highly reactive waste products. A radioactive waste can be a solid, a gaseous vapor or a liquid.

Uranium is used as a fuel for nuclear power plants. It is five percent enriched through a gaseous process by adding such things as radon, hexafluoride and thorium.

When uranium is burned, it causes a chain reaction type of energy. The burnt out uranium is called 'depleted uranium' (DU) and that is a highly radioactive waste product. It is 60 percent as radioactive as naturally occurring uranium and its half-life is four and a half billion years. $t_{1/2} = \frac{\ln(2)}{\lambda}$ This concoction of a calculation was discovered in 1907 and tells scientist and mathematicians just how long it takes a substance to decompose to half of its original weight. We just have to believe them. So, the fact is now known that half of the depleted uranium will still be around in four and a half billion years.

Currently the US has one billion pounds of DU in storage in an appropriately named location in Nevada - Yucca Mountain. That is the general consensus when considering having nuclear waste in one's back yard, yuck! It would be a reliable assumption that the people of Nevada don't want their state to be used as a landfill storage site for this highly toxic material. A large dose will kill many people while a small dose only results in dreaded diseases and devastating birth defects lasting decades in the animal kingdom including humans.

Co-Op America recently listed ten stunning reasons to avoid nuclear energy. Among the list is a growing concern that **living near a nuclear plant increases the risk for childhood leukemia and other forms of cancer** - even when a plant has an accident-free track record. One Texas study found increased cancer rates in north central Texas since the Comanche Peak nuclear power plant was established in 1990, and a recent German study found childhood leukemia clusters near several nuclear power sites in Europe. According to Dr. Helen Caldicott, a nuclear energy expert, nuclear power plants produce numerous dangerous, carcinogenic elements. Among them are:

- Iodine 131, which bio-concentrates in leafy vegetables and milk and can induce thyroid cancer
- Strontium 90, bio-concentrates in milk and bone, and can induce breast cancer, bone cancer, and leukemia.
- Cesium 137, which bio-concentrates in meat and can induce a malignant muscle cancer called a sarcoma
- Plutonium 239. A dangerous element that is carcinogenic in wee amounts as small as one-millionth of a gram and can cause liver cancer, bone cancer, lung cancer, testicular cancer, and birth defects.

http://www.coopamerica.org/programs/climate/dirtyenergy/nuclear.cfm

Congressional Research has found to date $66 billion of taxpayer dollars have been spent on research for nuclear energy. There are five states that have already shut

down some of their plants due to extreme cost (New York, Illinois, Oregon, Maine and Connecticut).
The potential for disaster, especially transporting uranium is of the highest concern. There have been 33 industrial transportations accidents since 1990. The most recent notable industrial chemical spill was Tripropylene in the Baltimore Tunnel rail fire July 18-23, 2001 when CSX Freight Train derailed. A fire ensued and burned at 1800 degrees for three days. What if this substance being transported was actually depleted uranium on its way to Nevada? Moving fuel makes an excellent target for terrorists, too.

Finally a plan to consolidate the wasted DU and other toxins into Plutonomium could fall into enemy arms and increase terrorism and support bomb productions. It may not be fossil fuel, but it is definitely not the best solution to our energy needs. We all need to know the facts before allowing one of these facilities to be built in our hometown.

Solar energy:

"I'd put my money on the sun and solar energy. What a source of power! I hope we don't have to wait until oil and coal run out before we tackle that."
Thomas Edison 1931

The US Department of Energy Efficiency and Renewable Energy authorities are encouraging Americans to make solar energy a conventional form of electricity by 2015 through Solar America Initiative. This will provide a clean source of power, improve our environment by decreasing 191,000 tons per year of CO_2 and will boost the economy by promoting the industry.

This initiative along with current tax incentives and rebates makes solar energy a very important goal and wise decision for a sustainable future.
Here's how it works: solar energy is absorbed in solar collectors to provide hot water or space heating in households and commercial buildings. Parabolic mirrors concentrate sunshine to provide heat greater than 1,000 Celsius. This heat can be used either for heating purposes or to generate electricity.
http://energy.saving.nu/solarenergy/energy.shtml

Another way to produce power from the sun is through photovoltaic methods. Solar Photovoltaic (PV) cells are devices that convert solar radiation directly into electricity. A Federal tax credit of 30 percent or $2,000 is available in addition to state incentives and rebates. PV systems require very little maintenance and are expected to last 30-40 years.

Here's what you need to get set up with PC:

- Southern roof exposure
- Bright sunlight for six hours per day
- Roof must be free of shade or shadows

- A minimum of 90 square feet is required for a six-panel system
- Roof should be less than ten years old
- No slate or cedar roofs

This is a lifetime savings benefit for you and the planet. Using just six panels prevents 2,500 pound of CO_2, about 6 pounds of Nitrous Oxide (NO_2) and over a dozen pounds of Sulfur dioxide (SO_2) every single year.

Catch a little sunbeam; put it in your battery and save it for a rainy day. For years, solar panels have been off-grid; that is to say, they were not connected to the electric company or meter from the home to the pole on the road. As power was generated, it was stored in batteries in the garage or basement. These homes were totally self-sufficient. However, in recent years, with the invention of effective "inverters" that interface between home and network make more and more Americans go to a "grid-tied" system. That is, when their homes generate power through the panels to collect energy. The house serves as a small utility company as the electricity is fed back into the utility company grid system. At night, people take power back from the network. Your meter will actually spin backward during the day! Your utility company will credit the power added to the system. Battery backup systems are available during power outages. But in essence, you use the grid as your giant storage battery. Sierra Magazine: http://www.sierraclub.org/sierra/200707/remodeling.asp

You're in hot water now: A solar hot-water system can supply 75 percent of the energy a family of four needs to heat its water. With photovoltaic panels, one can provide power with a passive solar collector. The average household hot water costs are 15-30 percent of the utility bill. The estimated payback period for a family-size system costing **$3,000 to $6,000** is ten to twenty years, but it should be shorter when you factor in tax credits and incentives.

World Watch Institute has released a report saying they expect the cost of solar panels to drop 40 percent in the next few years. The recent explosive growth in the solar power industry has caused a worldwide shortage in the process of producing solar panels. For more information, see builditsolar.com a comprehensive web site explains a variety of projects with step-by-step instructions for the do-it-yourselfer. Requirements are similar to the PV system above. However, a slightly larger roof size is necessary (10 x10) as well as a two 80- gallon water tanks. Solar hot water is even more effective than regular photovoltaic system. Not only are you cutting out 25 percent of your electric bill, solar heat prevents about 10 barrels of oil from entering power plants.

~GREEN BEINGS —SAVE GENES~

Solar technology on the 2010 horizon: The emerging technology uses so-called thin films mounted on glass windows and other surfaces to harness the sun's rays. It's more attractive and cheaper than the bulkier conventional solar cells. Many believe thin film's efficiency will steadily rise with improving technology. Taiwan's E-Ton Solar estimates thin film could account for up to 30 percent of the global solar cell market by 2010, up from around seven percent in 2006. Sharp Corp. plans to boost its thin film solar-cell production capacity from 15 megawatts a year to 1,000 megawatts by 2010 with the construction of a massive plant in Sakai City, Japan. http://www.latimes.com/business/la-ft-solar28jan28,1,1240524.story?ctrack=1&cset=true

Check out the green power in your state: This is from the Department of Energy. Discover energy efficient power in your state. Just click on your state and add your zip code. You are greener and wiser everyday.
http://www.eere.energy.gov/greenpower/buying/buying_power.shtml

It doesn't take Sherlock Holmes to uncover solar home locations. The winning solar cities named by US Department of Energy June 2007:
       1. Ann Arbor, MI
       2. Austin, TX
       3. Berkeley, CA

4. Boston, MA
5. Madison, WI
6. New Orleans, LA
7. New York City, NY
8. Pittsburgh, PA
9. Portland, OR
10. Salt Lake City, UT
11. San Diego, CA
12. San Francisco, CA
13. Tucson, AZ

If just 10 percent of New Yorkers purchased green power it would prevent nearly three billion pounds of CO_2 from entering the atmosphere each year. New Yorker can learn more about green power and renewable energy through Con Edison Solutions at 1-888-320-8991 or visit their web page. www.ConEdSolutions.com/greenpower

Take a look on the next page at the population in different areas of the country and see the average use of kilowatts per month. As you work your way through this book you will see how you compare to the national average. Pull that utility bill out again and keep on going with these tactics. Knowing your energy consumption helps to show you what your conservation efforts are actually doing to help keep money in your pocketbook and Mother Earth healthy.

Chart below is from Utili Point International
http://www.utilipoint.com/issuealert/article.asp?id=1728

Regions	Population	kWh
New England	5822935	618
Middle Atlantic	15045495	641
West North Central	8287837	903
South Atlantic	22473797	1088
East South Central	7356975	1193
Central West South Central	12883403	1151
Mountain	7368280	847
West Coast	15763570	668
Hawaii & Alaska	609661	642
US Total	114317707	877

For an extensive 600 page renewable energy bible consider the Real Goods Solar living Sourcebook 12th Edition http://www.gaiam.com/retail/product/21-0359

Wind power: Wind and solar power emit zero percent toxins to pollute the environment. They offer no support to terrorism nor are they harmful to humans, animals or the Earth. Wind generated electricity holds the greatest potential for renewable energy. At 13.9 cents per kilowatt the prices are about equal with coal-generated electricity.

The DOE reports the Great Plains are home to some of the most persistent wind in the world. In fact ND is capable of generating 20 percent of the US electricity! And if we consider the wind in the world just 20 percent of it can supply the entire planet seven times over. Another innovative planner comes from the Pickens Plan. www.pickensplan.com This recovered oil giant of a man is now using his power and wealth to develop wind solutions that will benefit all of humanity. We need more like him. Please visit his web to learn more.

Modern wind generators can produce over two megawatts of energy. It is possible that by 2030 wind will be able to produce over 250 giga-watts to replace 120 coal factories! The wind only needs to blow at 11 miles per hour for wind turbines to be cost effective. http://www.sierraclub.org/energy/slideshow/popointscript.pdf

BIG misconception; one does <u>not</u> need a windmill in their backyard or even their city to benefit from the energy supplied by wind. Electricity is generated on wind farms that go into the power company's grid from which we all pull our home's needs. When you purchase wind power you decrease the amount of fossil fuels used by your power company. Just as solar panels cause the electric meter in the home to spin backward, the wind power does the same at a larger station.

You can, however, own your own windmill. According to the American Wind and Energy Association, during 2006 there were 6,800 wind turbines sold for residential use in the US and in 2007 the number increased to 8,000 privately owned wind turbines. Currently the cost of purchasing and installing a residential wind turbine costs **$7,000-$12,000.** Yep, as Josh Donner of Sierra Club writes, "we need more windmills not windfall profits" as we have witness in the oil industry.

Rooftop wind turbines: Aerotecture International of Chicago has developed rooftop wind technology. This is a vertical axis wind turbine that can capture wind blowing from any direction. Instead of propellers mounted on tall poles, these turbines are curved, galvanized steel shaped like the double helix of DNA. Some of these turbines can generate electricity from an eight-mile-an-hour breeze to over 100-mile-per-hour gust to produce 2160-kilowatt hours of electricity. They are quiet, safe for wildlife, don't vibrate and costs start at **$3,000.**

Remember supply and demand. Demand renewable energy in the form of wind and solar energy. Supply will increase and naturally, gradually, the price will decline due to competitive marketplaces.

Geothermal: "Clean Sustainable Energy for the Benefit of Humanity and Environment" Wow, right from the horse's mouth...I mean the EPA's voice of authority. What is this geo thermal stuff anyway? This is the third most underutilized source of renewable energy. http://www.geothermal.org/GeoEnergy.pdf

It is important not to confuse a geothermal power plant with geothermal heat pumps. The geothermal power plant must drill a well down to a hot water reservoir deep in the earth. Reservoirs exist mainly in the western US, Alaska and Hawaii. Cold water is sent down and hot steaming water is pumped back up to the facility. This steam will provide rotational or mechanical energy to turn turbine blades. The energy from the turbine is used to spin magnets inside a coil to create electric current. That current is fed into power lines.

Ground source heat pump (GSHP): This type of geothermal heat pump is used in homes and business. It is quite a different set up and especially compatible for the eastern US. The usual temperature within ten feet of the Earth's surface is about 59 degrees, so a heat exchanger is natural. Similar to a cave, in winter or summer, the temperature is always the same.

The ground heat exchanger can be categorized by the different loops that are available. Pictured here are the closed loop pipes and the well water open loop. Again there are options for installation from: horizontal, vertical, pond/lake loop or even a slinky coil style for small spaces.

The most common (GSHP) system is the closed loop. It is made from high-density polyethylene pipe and is buried horizontally approximately five to twelve feet underground. Another type is vertical pipes descending between 100-400 hundred feet deep.

The pipes are filled with antifreeze. In the winter the solution in the pipe equalizes with the temperature of the Earth. This solution is returned to the home much warmer than when it departed the home. In this way the heat is exchanged, pumped back to the house or facility and then delivered into the air duct system. There is almost no need to change the house temperature during the summer. During the winter the home-air only needs to be warmed about ten degrees to reach 68-70 . In addition this process creates free hot water in the summer and delivers substantial hot water savings in the winter. What another nice gift from Mother Earth and a big savings for you!

http://www.igshpa.okstate.edu/geothermal/geothermal.htm

http://www1.eere.energy.gov/geothermal/gpp_animation.html

Dual source heat pump: According to The Department of Energy Efficiency and Renewable Energy, dual source heat pumps combine an air-source heat pump with a geothermal heat pump. These appliances combine the best of both worlds. Dual source pumps have higher efficiency ratings than air source units but are not as efficient as geothermal units. The main advantage is

they cost much less to install than a single geothermal unit and are almost as efficient.
http://www.eere.energy.gov/consumer/your_home/space_heating_cooling/index.cfm/mytopic=12640)

It is true the installation of geothermal can cost several times that of an air source system, but the savings will be appreciated after five to ten years. The system is expected to perform inside for 25 years and the outside components more than fifty years. There are about 40,000 geothermal pumps in the US. To find a qualified installer, call your local utility company or contact the International Ground Source Heat Pump Association or Geothermal Heat Pump Consortium for a list of certified and experienced installers. Checking references is always a good habit and can end up saving you even more when you know what type of problems have previously been encountered.

Chapter Fourteen

Empowering your Transportation

You better start this section with a dose of the prescription medicine Antivert. If your head wasn't spinning before, this next part will give that person with the strongest equilibrium a huge case of vertigo and gives *The Good the Bad and the Ugly* a whole new meaning.

Biofuel can be manufactured from any living organic material. As you now know, the ancient skeletal fossil remains of animal and plant life resulted in the by-products of coal, oil and gas over millions of years. Today, we also know how to recover the carbon from recently living plants and we know that fermenting produces ethanol from starch and sugar crops such as corn, soybeans, grasses, sugar, algae, manure or just plain old garbage. Check out the alternative fuel locator at this site. http://journeytoforever.org/ethanol.html http://www.eere.energy.gov/afdc/fuels/stations_locator.html

Henry Ford designed the Model T to run on ethanol (ethyl alcohol, grain alcohol), describing in 1925, the fuel of the future.
 "The fuel of the future is going to come from fruit like that sumac out by the road, or from apples, weeds, sawdust - almost anything. There is fuel in every bit of vegetable matter that can be fermented. There's enough

alcohol in one year's yield of an acre of potatoes to drive the machinery necessary to cultivate the fields for a hundred years"

http://www.gminsidenews.com/forums/f19/rudolf-diesel-henry-ford-ford-quotes-bio-fuels-53144

Here we are decades later and the race is on to return to efficient and economical fuel for the advantage of the people, the planet and not the oil industry. Henry was ingenious, but it has taken a long time to follow his design. Today the US is producing about 15 billion gallons of ethanol each year (12 percent of all fuel) and Brazil is producing and using 24 percent ethanol. Still it may not be the magic bullet we have been seeking. In thinking about corn ethanol, here are some considerations.

Proponents' arguments for corn ethanol:

- Cleaner fuel than gasoline
- Renewable fuel made from plants
- Manufacturing it and burning it does not increase pollution
- Provides high octane at low cost
- Can be used in all petrol engines without modifications
- Biodegradable without harmful effects
- Reduces harmful exhaust emissions
- Reduces carbon monoxide levels by 25-30 percent
- Reduces emissions of hydrocarbons

- Reduces nitrogen oxide emissions by up to 20 percent
- Reduces emissions of VOCs by 30 percent
- Cuts emissions of cancer-causing benzene and butadiene by more than 50 percent
- Reduces emission of SO_2 and Particulate Matter (PM)

Read more about Ethanol Fuel in the Governors' Ethanol Coalition (http://www.ethanol-gec.org/pub.htm), an organization dedicated to increasing the use of ethanol-based fuels.

Too good to be true? Then, it probably isn't. There are definitely drawbacks to using only **corn** ethanol as the magic fuel. It seems we are putting up a lot of bucks and not getting as much bang. This is startling.

If the current gasoline engine required a mere three percent improvement in fuel economy, that would surpass the entire 2006 corn ethanol production, Wow, falling back to efficiency again and again. In addition, corn has a use already as food for animals and humans. Corn requires intensive cultivation, fertilization, water, and energy.

Furthermore, the landmass that would be necessary to fuel just five percent of the US licensed drivers' needs would be massive covering several states entirely. This would obviously destroy thousands of habitats for all life forms including maybe your own. So now you know corn ethanol is not the answer, either.

There is something even better on the horizon, but it's not yet available. New research shows that cellulosic ethanol produced from switch grass yields twice that of corn. This product would decrease global warming by a whopping 90 percent and is much closer to solving the problems we face today.

What about biodiesel? This clean-burning alternative fuel contains no petroleum. Mixing fatty substances such as soybean oil with an alcohol and an enzyme to produce a biodegradable nontoxic fuel produces biodiesel. Biodiesel can be used in diesel engines without much modification. If it is mixed with petroleum, it is labeled BXX where XX represents 20-80 percent petroleum. This is the only fuel to have passed the 1990 Clear Air Amendment and it's registered with the Environmental Protection Agency. When biodiesel fuel is burned, it is absorbed by plants and releases few if any toxins into the environment thereby reducing global warming by 50 percent.

Find a distributor near you: Call 800-841-5849 www.biodiesel.org or http://journeytoforever.org/biodiesel_link.html, or the National Biodiesel Board at: http://www.biodiesel.org/resources/fuelfactsheets for the clearest chart in the literature, visit http://www.sierraclub.org/sierra/200709/biofuelschart.pdf to help you sort out all the pros and cons of fuels.

Dirty tricks: While petroleum diesels are more efficient than gasoline-powered vehicles, they produce higher levels of environmentally damaging nitrogen oxides and particulate matter and consequently require more sophisticated emission controls. Nitrogen oxides, or NOx, are the generic term for a group of highly reactive gases, all of which contain nitrogen and oxygen in varying amounts. Many of the nitrogen oxides are colorless and odorless. Nitrogen oxides form when fuel is burned at high temperatures, as in a combustion process. The primary sources of NOx are motor vehicles, electric utilities, and other industrial, commercial, and residential sources that burn fuels. http://www.envirotools.org/factsheets/contaminants/nitrogenoxides.shtml

Diesel is a dirty fuel costing an average of $4.29 per gallon (5/08). The EPA estimates there are 11 million older diesel engines in the US lacking emission control technology that produce more than 1,000 tons of particulate matter pollution every day and cause approximately 21,000 premature deaths in the US alone each year, many due to respiratory ailments. Rep. John D. Dingell, D-Dearborn, chairman of the House Energy and Commerce Committee has introduced a bill to the senate to retrofit these old engines making them less of a pollution problem to the environment.

WE ARE TRYING TO MODIFY OUR MODELS FOR USE WHEN FOSSIL FUEL IS DEPLETED.

Good news/ bad news: It is true that diesel fuel engines are dirtier and more harmful to the environment. But, the good news is a diesel engine can be a retrofitted or converted to run on used french fry oil (an environmentally friendly use for Mac Donald's after all) or any vegetable oil for around $1,000. Commercial trucks can be converted for about $10,000. Yep, this is the same vegetable oil that most restaurants are paying to dump or have hauled away in their trash. And most restaurants would give it to you for free or even pay you to remove it for them. Can you imagine that? Free fuel! Now that is good news with a laugh!

Don't be fossil fools: Besides the environmental harm petroleum causes to life on the planet, need we even mention the extreme price to our pocketbook? Maui is the first place in the US to top $4 a gallon for regular gasoline. The remainder of US currently averages $3.89 per gallon. Crude oil is now at $129 a barrel and rising about a dollar a day for the last several weeks. Seems they will keep raising it as long as we are willing to pay it.

Think alternative: carpool if you must continue commuting in a car. This is a way to cut your expenses in half or third immediately. As Mr. Gore reminded us, every pound of coal... every gallon of gasoline or oil... every unit

~GREEN BEINGS —SAVE GENES~

of electricity... every fossil fuel we burn... every aspect of our lives has some environmental impact.

On the horizon: It's a little like putting the cart before the horse, but White Plains, NY has the first hydrogen refueling station in the US for public use. Hydrogen-powered vehicles produce no carbon emissions. In fact, only water vapor comes from the exhaust pipe. Mass production of an affordable hydrogen fuel cell car is only a short time away.

However, in the spring of 2008, five families were selected to receive specially equipped hydrogen powered General Motors' vehicles costing $90,000 each. The research program, called Project Driveway, can be followed on the web. This is a three-year program to evaluate consumer opinion. Do we really need three years to voice our opinion? http://www.chevrolet.com/fuelcell

Chapter Fifteen

Second Biggest Investment
Thy Car

American cars, light trucks and their fuel: According to Alliance to Save Energy and US Department of Energy, American autos use over eight million barrels of oil a day. An average car emits 35 pounds of carbon dioxide every day. Over the course of one year with an average of 15,000 miles driven, most automobiles emit almost 12,000 pounds of CO_2. Drive a fuel-efficient car and save about 8,000 pounds CO_2 per year.

A car that gets 30 mpg will emit about half the CO_2 of a 15-mpg vehicle. That's a savings of 8,000 pounds if you drive 12,000 miles a year. In one year, an acre of trees absorbs the amount of CO_2 produced from driving a car 26,000 miles.

America consumes over 25 percent of the world's oil production. That boils down to 17 million barrels of oil per day. Sixty-six percent of oil (8 million barrels) is used for transportation for our beloved vehicles. Forty percent of US oil comes from foreign ports.

~GREEN BEINGS —SAVE GENES~

"SUPPORTING FUEL ECONOMY
WOULD JUST DEVASTATE THE
POCKETS OF SO MANY OIL
INVESTORS AND OWNERS"

Oil and natural gas, like coal, (discussed Chapter Thirteen) are products resulting from 500 million years of ancient plant and animal matter being buried under stone or mud. As organisms became trapped under high pressure without oxygen, bacteria decomposed them and oil was eventually formed. In some parts of the Earth we must drill to find oil. In other places such as Iran and Kuwait, the oil has been known to just seep out of the ground. Even so, worldwide reserves could be depleted in 40 to 60 years at the current rate of consumption.
http://www.petroleumequities.com/OilSupplyReport.htm

We must find alternative sources of fuel to run our cars and produce heat, not just in fear of running out of oil; but to curb carbon pollution that is harming the planet and us. Burning fossil fuels is one of the main causes of excess greenhouse gas resulting in global climate change.

Drilling and manufacturing oil has proven to be an environmental disaster that has contaminated our air, land, water and marine wildlife, too. Between 1973 and 1993 200,000 oil spills occurred worldwide dumping 230 million gallons of oil in to the world's waterways. We've all seen the pictures of birds with feathers drenched in oil, unable to swim, and drowning as well as uncounted fish and coral reefs dying from exposure to oil spills.

Many people are familiar with the Exxon Valdez 11 million gallon spill that covered 1,300 square miles of water and shoreline. "By comparison, oil intentionally released from Kuwaiti refineries and terminals by Iraqi troops amounted to 250 million gallons. In addition, they lit more than 700 oil wells, putting hundreds of tons of smoke and toxic chemicals into the air. In a list of the world's largest oil spills between 1967 and 1992, the Exxon Valdez spill ranks as number 36.
http://www.ucsUS.org/clean_energy/fossil_fuels/offmen-how-oil-works.html

The US Government, along with automakers, developed incentives to stimulate consumers to purchase fuel-efficient automobiles. These credits are available for vehicles that meet strict criteria for a limited amount of time. The current credits last only until 2010. One of the criteria automakers must fulfill is a standard of fuel economy. The minimum standard must be 25 percent over a baseline vehicle of 2002. Each 25 percent improvement over baseline improves the tax credit to a maximum of 250 percent or a dollar value of $400. There are additional conservation credits when a vehicle saves between 1,200-3,000 gallons of fuel over its expected lifetime. Combining both rebates then brings the incentive to $3,400.

However, there always seems to be a catch. After a manufacturer sells 60,000 qualifying vehicles, the tax credit is phased out over a period of fifteen months for vehicles that the manufacturer produces, which is why Toyota rebates are no longer available on Prius or Lexus hybrids purchased after September 30, 2007.
Find the best and worst automobiles listed by Greener Cars dot com. Some states have passed laws regulating auto emissions that are stricter than the Federal government. At this writing the EPA has denied the necessary waiver to California, required to allow better control of our environment through fuel standards.
http://www.greenercars.com American Council for Energy Efficient Economy http://www.aceee.org/

http://www.odemagazine.com/doc/51/greener-than-a-hybrid/

Here are even some newer models not yet on the list:
- Volkswagen's new Polo Blue Motion—a low-weight vehicle with a three-cylinder diesel engine—emits less carbon than the Prius and gets slightly better gas mileage, (99 to 104 grams of CO_2 per kilometer traveled, compared to 106 for the Prius).
- The Peugeot 107 grams of CO_2
- Citroën
- Toyota Aygo also comes close with 109 grams of CO_2 per kilometer
- Mazda 2 hatchback, soon to go on sale, also emits 109 gm CO_2

On average, every new SUV cruising the streets produces 60 percent more climate-threatening CO_2 emissions than a new car. And then there are efficient models such as the Honda Insight. The Insight cuts emissions 50 percent in comparison to similar autos. A smaller car will have better efficiency. There are claims that driving an Insight instead of a Mini Cooper decreases CO_2 emissions by 2.9 tons over 15,000 miles of annual driving. And at the other extreme, driving a Hummer H2 instead of an average car will increase annual CO_2 emissions by 5.2 tons. That means we need two Honda Insight autos to negate one un-harmful Hummer driver. With what we know today, vehicles can be fuel-efficient. Fuel economy should be the number one consideration, followed by

~GREEN BEINGS —SAVE GENES~

emissions and carbon footprint when purchasing a new vehicle. Things such as price, color, power and other features need to take their place in the back seat. Auto price, resale price, emissions and fuel economy are considerations that should replace all other features.
http://www.greenercars.com/highlights_meanest.htm

Worst New Vehicles for the Environment in 2008

Make and Model	MPG: City	MPG: Hwy	Green Score
BUGATTI VEYRON	8	13	8
LAMBORGHINI MURCIELAGO /	8	13	8
MERCEDES-BENZ G55 AMG	11	13	11
MURCIELAGO ROADSTER	8	13	8
BENTLEY ARNAGE d	9	15	9
BENTLEY AZURE	9	15	9
GMC YUKON 2500 c	12	16	12
HUMMER H2 c	12	16	12
VOLKSWAGEN TOUAREG	15	20	15
JEEP GRAND CHEROKEE	17	22	17
MERCEDES-BENZ GL320 CDI	18	24	18
MERCEDES-BENZ ML320 CDI	18	24	18
MERCEDES-BENZ R320 CDI	18	24	18

Available to subscribers of the *ACEEE's Green Book® Online* interactive database are summary "Green Scores" of the 1,300+ configurations of all model year 2008 vehicles, along with:

- Each car's configured fuel economy, both city and highway
- Health-related pollution impacts
- Global warming emissions
- Estimated fuel expenses

It is important to note that though some vehicles appear more efficient due to their fuel consumption, it is the entire picture that accounts for the green score. This score reflects the car's pollution level, amount of damage to the environment creating the vehicle as well as its ability to be recycled. Subscribers to *ACEEE's Green Book® Online* can also build custom lists for comparing vehicles. Monthly and annual subscriptions to *ACEEE's Green Book® Online* is available at GreenerCars.com

Best New Vehicles for the Environment in 2008

Make and Model	MPG: City	MPG: Hwy	Green Score
TOYOTA PRIUS	48	45	53
HONDA CIVIC HYBRID	40	45	51
SMART FORTWO CONVERTIBLE/COUPE	33	41	49
HONDA FIT	27	34	43
FORD ESCAPE HYBRID	34	30	42
HYUNDAI SONATA	21	30	39
SUBARU OUTBACK WAGON	20	26	37
NISSAN ROGUE	22	27	37
TOYOTA TACOMA	19	25	34
TOYOTA SIENNA	17	23	33
CHEVROLET TAHOE HYBRID C1500	21	22	28
NISSAN FRONTIER	14	19	27

Press materials are available from ACEEE Publications. For further information, contact:

ACEEE Publications
1001 Connecticut Avenue, NW, Suite 801
Washington, DC 20036-5525
Phone: 202-429-0063
Fax: 202-429-0193
Email: aceee_publications@aceee.org
Web site: www.aceee.org

The hybrid-electric vehicle (HEV): Save money, save the planet and drive in ultimate style. Tesla Motors has the best sports car available and it is a 100 percent electric car that can go zero-60 mph in four seconds. The battery requires charging every 245 miles, similar to the gasoline fill up for many guzzling SUV's out there. The cost to operate the Tesla is about two cents per mile. If that were fossil fuel it would be like getting 135 miles to the gallon. Unfortunately, at the moment the cost of this particular vehicle is out of range for all but two to three percent of the population. Take a look though to see the future. Tesla offers proof that efficiency can be had and placed their proof in a very desirable vehicle. http://www.teslamotors.com/

Hot off the press is another plug-in vehicle that is caught up in a legal battle with Tesla over design infringement. Let's hope that strife can be settled for a new company White Star, for their perfectly named rival Karma. Until then, the most affordable and available option is Toyota's Prius, a great example of hybrid technology. The average Prius cost about $23,000. Depending on the cost of gasoline, one can expect to save about $800 dollars per year and spare the planet an additional 4,000 pounds of carbon pollution.

Plug-in vehicle (PIV): What Happened to the Electric Car? It's on the way back, with your help, of course. If the plug-in vehicle is powered from electricity produced from wind or solar, it becomes 100 percent emission free preventing air pollution completely. If powered with electricity from a coal-fired power plant, the car is still 70 percent more effective than petroleum alone. Electric vehicles will help to reduce or eliminate our dependence on mid-east foreign oil suppliers. The Electric Automotive Association http://www.eaaev.org and **Plug in Partners** http://www.pluginpartners.org

Emission Comparisons:
Wind-driven Electric Gallon of Gasoline

POLLUTANT	ELECTRIC	GASOLINE
Nitrogen Oxide	0 pounds	41 pounds
Carbon Dioxide	0 pounds	10,000 pounds
Hydrocarbons	0 pounds	80 pounds
Carbon Monoxide	0 pounds	606 pounds
Sulfur Dioxide	0 pounds	trace amounts
Mercury	0 amounts	trace amounts

Table above is data based on 12,500 miles/year.

There are currently 16 different electric autos available with prices as low as $11,000-14,000 for example, the new US Smart for Two.

At 106 inches long, it is very similar to the East Indian Nano. This smarty has the power of 71 horses and is rated at 36 miles per gallon. The Nano will sell for quite a bit less- $2,500. Nope, it is not a misprint.

The 4-door Nano is ten feet long and five feet wide. It can reach 75 mph with its two-cycle 35-horse power engine, but is not available in America yet. Both Smart for Two and Nano have small trunks and are a perfect solution for travel around and about town. Change is happening, but sometimes it is not quick enough. As one stumbling block is conquered another seems to surface. US auto manufacturers claim plug-in-vehicles (PIV's) will cost an additional $2,000 to 3,000 dollars over that of the standard petroleum-burning engines. The plug-in battery range is expected to provide service for approximately 20-25 miles. The Electric Power Research Institute claims that 73 percent of autos on the road commute 25 miles or less to employment locations each day. Commuters traveling this distance or less would certainly benefit from plug in hybrid electric vehicle.

The batteries are promised to be even more efficient in hybrid vehicles (using fuel and battery) in the near future. Total number of hybrids on the road was roughly 620,000 at the end of 2006. The market, which has already branched into most major vehicle classes, received another couple of high-profile nameplates this model year in the Toyota Camry Hybrid and Nissan Altima Hybrid, each averaging 39 miles per gallon in combined city/highway driving.

The future is on the way with Honda: In 2003 Honda introduced the Experimental Home Energy Station IV. This appliance-like device may help solve problems of hydrogen supply for the zero emission fuel cell vehicles FCX Clarity that also uses a home's natural gas to produce hydrogen for heat and electricity. This fourth-generation fueling station is in Torrance, California. It is estimated that this combination can reduce CO2 emissions by an estimated 30 percent and energy costs by an estimated 50 percent. Hopefully the next generation designed will be even more energy efficient. They are on the right track.

http://world.honda.com/news/2007/4071114Experimental-Home-Energy-Station/index.html

Take the old dog back: Again, we must solute California for their initiative to lead another great program. When you learn that your old car is indeed a horrid polluter and asthma inducers, if you live in California, they will buy the auto back from you. What a great incentive to rid ourselves of these powerful pollutants on wheels. The voluntary accelerated vehicle retirement (VAVR) program or "old vehicle buy back program" provides monetary or other incentives to vehicle owners to voluntarily retire their older cars. A primary goal of the VAVR program is to encourage a more timely removal of older, polluting vehicles from California roadways to be replaced with newer, cleaner, more efficient vehicles or alternative transportation options. Write your congressman and request that your state offer a similar program. Environmental Working Group: http://www.ewg.org

So you want to be a zipster? What is a zipster? It is a person that does not want to own a personal automobile but occasionally requires the use of one. It is the latest for big-city dwellers that can use mass transit almost everywhere they go. With a zip car it is easy to rent a car in most cities, by the hour for about $10. The occasional zip-car driver pays a $50 annual fee or a $25 one-time application fee for vehicles that cost between nine and eleven dollars per hour or $66 to $77 per day. The great thing is this includes your fuel, insurance, maintenance and reserved parking in selected spots. More frequent users pay no annual fee and do receive a discounted hourly or daily rate.
http://www.zipcar.com/find-cars

Alternative to driving: Green machine bamboo-bicycle! Now, this product should receive double green-stamps. Not only is it made from a renewable resource, it requires no petroleum for transportation. There are some real geniuses out there! Riding a bike can easily save you over $4500 per year and again spare the air another 2800 pounds of carbon dioxide.
http://www.calfeedesign.com/bamboo.htm

Good Old Shoe Leather: Yes, if we all walked more, the planet would certainly benefit and you, as a smarty, know your health would too. Check out this cool site. I was amazed by the facts. Find out what is within walking distance of your own home. You will really be surprised by what is around the corner. Get yourself a pair of trekking sticks. They allow you to walk father with less effort. And they are pretty good at warding off evil dogs. Most fuel consumption is used for quick errands close to home. This is another way to save money and promote a healthier lifestyle.
http://www.walkscore.com

Chapter Sixteen

The Abode
No Pain ~ No Gain

My home is my castle? Perhaps soon even you may claim your home as a mini power plant. Americans have been building castles. With the right design a preplanned home of modest size will be more comfortable, cost less, and meet the needs and desires of even the most selective consumer.

Your biggest investment is still your home. Why not let yours pay you back by incorporating innovative technology and energy conservation methods? There are several ways: from Energy Star products to geo-thermal heating and cooling to solar roofs from being off the grid or even grid-tied to other power sources. Here are some thoughts and considerations.

A private home: Did you know...the average home releases more than 24,000 pounds of CO_2 annually, almost twice as much as a typical car, estimates the Environmental Protection Agency. This is due to emissions produced by power plants to generate the electricity used to run modern homes — plus home emissions from such things as oil and gas-fired furnaces.

A home constructed around energy efficiency can realize enormous savings. At the onset, positioning the house on the lot properly allows for best use of daylight and natural ventilation. Usually, a home facing south offers the most benefit.

The heating cooling, lighting choices you make for appliances and renewable energy systems can push a building closer and closer to net zero energy consumption. If you are considering building a home or are ready to do some serious renovations, be aware that energy efficiency is a key design criterion to increase the value and comfort of your home. The Energy Star rating system has a home certification program. This site will show you the 'must have' qualifications to be rated energy efficient.

http://www.energystar.gov/index.cfm?c=behind_the_walls.btw_landing
http://www.energystar.gov/index.cfm?c=new_homes.hm_index
http://www.eeba.org

"Permit for Pollution Please"

In addition, the US Green Building Council has created guidelines to achieve energy efficiency and sustainability called LEED (Leadership in Energy and Environmental Design)
http://www.usgbc.org/DisplayPage.aspx?CategoryID=19

This is a rating system for residential homes that focuses on new construction. It is estimated to cost about four to five percent more to build within the framework of LEED than with conventional construction methods. LEED home rewards includes economic benefits such as lower energy and water bills, environmental benefits like reduced greenhouse gas emissions, and health benefits such as reduced exposure to mold, mildew, indoor toxins and increases the overall appraised value of your home. Even better, the net cost of building a LEED home is comparable to that of building a conventional home. Contact LEED Customer Service at 1-800-795-1747 or leedinfo@usgbc.org

"Learn from yesterday, live for today, hope for tomorrow"

Albert Einstein

Residential Energy Service Network: (RESNET) is another not-for-profit national industry that offers building certifications. RESNET is recognized by:

- Mortgage industry for capitalizing energy efficiency in mortgages
- Financial industry for certification
- Federal government for verification of building energy performance
- Federal tax credit qualification, EPA ENERGY STAR labeled homes, US Department of Energy Building America program and States for minimum code compliance in 16 states.

http://www.natresnet.org/about/resnet.htm

Passive solar: Passive solar systems in a home utilize materials and design elements to capture the sun's energy and move that heat throughout the house. Some examples of passive solar design are:

- Utilizing overhangs to shade a house during the heat of the summer and allow sunlight to penetrate the interior of the house during the winter;

- South-facing windows with few windows on the west;

- Thermal absorption mass and techniques;

- Cross ventilation;

- Angle and positioning of the house on the property;

- Appropriate roof pitch;

- Horizontal shutters.

These techniques can reduce or even eliminate the need for air conditioning in homes. For example, with a well-designed overhang, with south-facing windows that admit the low-angled rays of the winter sun are shaded from the high-angled summer sun. Thermal mass such as a stone fireplace that receives and stores heat in the winter to release in the evening, works in reverse in the summer. The mass cools down in the evening and retains that coolness the next day, moderating the effects of high daytime temperatures.
http://energy.saving.nu/solarenergy/energy.shtml

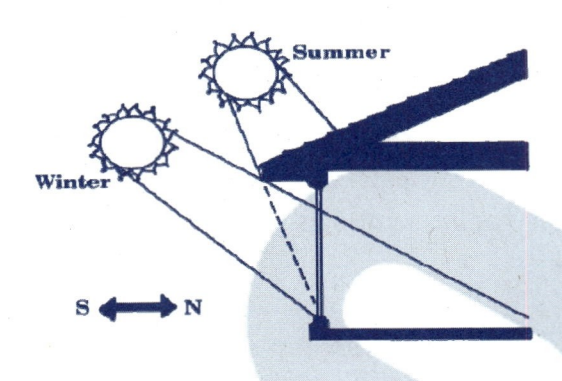

Graphic courtesy of North Carolina Solar Center

L.E.A.F. house: This winner of the Solar Decathlon is a home design with the latest technology coupled with the smartest resource efficiency. LEAF stands for Leading Everyone to an Abundant Future. The design comes from the University of Maryland and has integrated the latest innovative designs winning several awards in national competitions. Visit their site and get information to build it yourself. http://solarteam.org/page.php?id=250
Blue prints for a healthy home: Green Communities now offers Community Tree Grants in partnership with The Home Depot Foundation, to affordable housing developers to strategically incorporate trees into their site plans.
http://www.greencommunitiesonline.org/community_trees

Green roofs: Green roofs don't refer to the color but to the techniques of growing plants on the roof. This takes special preparation but if you are about to build or replace a roof contact the Green Roof Association. Plants growing on a roof keep the house cooler, prevent polluted water run off, and prevent soil erosion and water contamination. Green roofs also benefit solar panels...actually allowing them to be more efficient. Before you build or if your roof needs replacing please consider this high-tech energy saving tactic. With the proper construction, shape and strength you can have a beautiful roof covered with plants, flowers and wild grasses. No need to take the lawn mower up there, though: this common sense practice has been done for centuries and the old customs are finding their way back to the minds of people all over the world. Here are a few more reasons:

- **Lower energy costs:** Soil and plants add an extra layer of insulation to your home, keeping it cooler in summer and warmer in winter. This is particularly helpful in cities, where pavement and buildings reflect heat and raise air temperatures about 10 degrees.
- **Environmental benefits:** Rooftop vegetation provides food and shelter for insects, birds, and other wildlife, and plants' natural mechanisms for filtering impurities help improve air and water quality. Green roofs can also play a small role in flood prevention by reducing storm runoff.

- **A longer-lasting roof**: By providing a buffer against temperature extremes, wind, and heavy rain, soil and plants can help protect the underlying roof and extend its useful lifetime.
http://www.greenroofs.net/index.php?option=com_content&task=view&id=26&Itemid=40
http://www.greenroofs.com

If this is beyond your construction capabilities, please try another efficient option but avoid the use of fossil fuel-derived asphalt shingles. As mentioned in Chapter One, even a white painted tin roof is better than asphalt shingles.

Solar shingles: Don't waste sunshine, put your roof to work. Photovoltaic (PV) shingles provide the same look, protection, and durability of asphalt shingles but have the added benefit of converting sunlight into electricity to power your home or specific appliances, reducing the need for electricity generated from fossil fuels and lowering your electricity costs.

PV shingles work best on south-facing roofs that are not shaded by trees for a significant portion of the day. **Costs**: These environmentally friendly roofing options do cost more up front: approximately $10 to $25 per square foot (including installation) for green roofs and $20 per square foot (materials only) for PV shingles, compared with $2 a square foot for asphalt shingles. Unlike asphalt shingles, however, green roofs and PV shingles can save energy (and money) over their lifetime.

Chapter Seventeen

Odds for the End

Big foot, check your carbon footprint: Calculate your impact on the world today. Compare what you've accomplished already by altering your old habits to those of an energy-efficient responsible citizen of America willing to join the band. See what you can do for the next generation. *Will-Power* is a renewable resource.
http://www.climatecrisis.net/takeaction/carboncalculator/#
US Environmental Protection Agency's personal emission calculator
http://www.epa.gov/climatechange/emissions/ind_calculator.html
http://www.nativeenergy.com/pages/co_op_america/154.php?afc=Coop

Conserve what you can, offset what you can't.

reduce what you can, offset what you can't
- driving
- consumables
- home energy
- transportation
- flying

offset
- shared services

http://brighterplanet.com/knowledgebase/conserve_and_offset

Here's how: Brighter Planet and Bank of America (BOA) have teamed up to help you reduce your carbon footprint.

Earn 1 Earth-Smart™ point for every $1 spent in net retail purchases. Points are automatically redeemed monthly to help fund renewable energy projects. Every 1,000 points will fund an estimated 1-ton of carbon offsets. Every 1,000 points is roughly equivalent to taking a car off the road for 2,000 miles, or powering and -heating/cooling your home for a month. Earn 50 percent more points with a match from Bank of America through December 2008. After your first transaction earn 1,000 bonus points to fund an estimated one-ton of carbon offsets to offset the creation and delivery of your credit card. Earn an additional 1,000 bonus points when you sign up for paperless statements.
bankofamerica.com/environment

Green washing: True colors come through; there is another side to the BOA green promotion. It is the leading financial institution supporting and funding new coal-fired power plants as well and the related destructive mountain top removal. Ken Lewis, Bank of America's chairman, president and CEO was awarded the biggest Fossil Fool of the year (2008), a recognition given to the world's biggest contributor to our global addiction to fossil fuels. This is more evidence that every action has an opposite and equal reaction.

Use your enlightened energy to write a letter to BOA and voice your disapproval of their use of your invested money or considering moving your savings to another institute. Coop America:
http://www.coopamerica.org/cabn/newsletter/announcements/200804/index.cfm

Need a job? Over 4,000 PV installer jobs in the US are available. Currently, there are about 300 certified installers of solar photovoltaic (PV) panels by the North American Board of Certified Energy Professionals (NABCEP). However, The US Department of Energy's (DOE's) Energy Efficiency and Renewable Energy Office estimate the US needs 4,700 certified installers. For buyers to receive the promised incentives, PV installers must be "master electricians who have completed a training course to prepare for NABCEP certification or who work with NABCEP-certified electricians."
http://www.nabcep.org/pv_installer.cfm

The Maine Public Utilities Commission, on the other hand, must qualify solar thermal installers. As of April 2008, 513 individuals have passed the NABCEP certification as qualified PV installers. Many are certified electricians. There are more qualified solar thermal installers. The higher number of qualified thermal installers most likely reflects the ease of getting that license. A solar thermal installer may obtain a license based on experience alone without a formal education, though many different levels of education are also accepted as a prerequisite to licensing. For more information see the Journal Distributed Energy
http://www.distributedenergy.com/de_0705_sunny.html

Shift your purchases to Green Businesses: Consumers and investors can send a strong message to businesses by shifting their spending and investing to green and Fair Trade businesses committed to helping-not harming, people and the planet. This is the most powerful way to grow a just and environmentally sustainable marketplace. Many businesses are now being held accountable for the life of their products. This responsibility will go beyond the production of consumer goods but also includes the destruction or the recycling of the same products. As the manufacturer becomes fully responsible for their product a new consciousness will be realized. Support companies that offer these additional green benefits.

Global markets for biofuels, wind power, solar photovoltaic, and fuel cells jumped 40 percent last year to a total of $77.3 billion. If the current growth rates continues, these four key sectors are expected to more than triple over the next ten years to a total of $254.5 billion by 2017, according to a new report from research firm Clean Edge.
http://www.cleanedge.com/reports/charts-reports-trends2008.php
Venture capital investment in clean technologies climbed 43 percent last year, according Dow Jones Venture Source.

Clean-Energy Venture Capital Investments in U.S.-Based Companies as Percent of Total

Year	Total Venture Investments (US$ Billions)	Energy Technology Investments (US$ Millions)	Energy Technology Percentage of Venture Total
2000	$105.1	$599	0.6%
2001	$40.6	$584	1.4%
2002	$22	$483	2.2%
2003	$19.7	$446	2.3%
2004	$22.5	$663	2.9%
2005	$23	$1,038	4.5%
2006	$26.5	$1,555	5.9%
2007	$29.4	$2,665	9.1%

Source: New Energy Finance with supporting data from Nth Power and Clean Edge. NOTE: New Energy Finance's energy-tech VC numbers include investment in renewable energy, biofuels, low-carbon technologies, and the carbon markets. VC figures are for development and initial commercialization of technologies, products and services, and do not include private investments in public equity (PIPE) or expansion capital deals.

U.S. Top 10 Disclosed Energy-Tech Venture Deals (2007)

Company	Primary sector	Total invested (U.S. $ Millions)
HelioVolt Corporation	Solar	$100.5
GreatPoint Energy	Efficiency: Supply Side	$100.0
Arcadian Networks	Efficiency: Supply Side	$90.0
Solyndra Inc	Solar	$79.2
SolFocus Inc	Solar	$63.6
Calera Corporation	CCS	$58.5
Miasolé Inc	Solar	$50.0
Solaria Corp	Solar	$50.0

Source: New Energy Finance, 2008

http://www.sustainablelifemedia.com/content/story/climate/clean_energy_market_to_triple_within_ten_years

Chapter Eighteen

Conclusion

So, there you have it folks-conservation in a nutshell. You have just been given hundreds of ways to make a difference. You have learned techniques to help you save money. You have been given suggestions and ideas for the best places or products to spend your hard earned salary and just as importantly the toxins and pollutants to boycott indefinitely.

Through all-these methods you are aware of your role in protecting the planet, decreasing carbon emissions below **350** ppm, and leaving a better home for future generations. You have learned how to make informed decisions to protect yourself, family and the environment. You have wisdom for best tactic of all- efficiency. You are more aware of your impact and personal responsibility in the world in which we live. You expect your government and businesses to live up to the same responsibilities and have a right to say so.

Contact your local political representative. Ask what they are doing to conserve energy. Voice your opinion and remind them that your vote will go to the best candidate who will guide us to clean sustainable energy, a healthy environmental, justice and renewable methods of production and disposal.

We must participate in the political process to ensure an energy efficient future. The squeaky door gets the grease. Squeak out!

One of the most important things you can do is call or to write your representatives at all level of government and let him or her know your thoughts on pending legislation. To identifying your representative in the US House of Representatives visit the web site below.
http://www.house.gov/writerep or call Washington, DC 20515 (202) 224-3121

Go figure: Here is a mind-boggling math problem to leave you thinking. It is another version of the "cap and trade" scheme that the political system is currently trying to design.
The goal is to equally share the carbon dioxide burden of our country. So, it seems we also should be equally awarded individual carbon credits for our efficient efforts. We can then sell our excess credits back to US corporations that pollute the air. Here are some facts and figures:

- The US discharged 4,990.6 million metric tons of CO_2 into the atmosphere in 1990:
- The goal is to decrease our global carbon emission to **350** ppm. That amount would equal 3743 million metric tons of CO_2 by 2012.
- There is one ton of carbon in 3.67 tons of CO_2

~GREEN BEINGS —SAVE GENES~

- ◆ The current population of the US is 301,139,947:
- ◆ One ton of carbon is a credit worth $10 (in US currency):
- ◆ How many carbon credits would be allocated for every single person in the US?
- ◆ How many dollars per month would every person receive for his or her carbon allowance?

Suddenly, each individual and family has an incentive to become energy efficient, to save their carbon allowance. It is another source of income to sell your carbon credits. Your carbon payment would be required to purchase your automobile gasoline as well as pay your utility bills. This payment is not instead of cash or credit card payment but it gives you an allotment of carbon pollution in the world. Remember if you are using wind, solar or any renewable energy source, you obviously would not be spending any of your carbon credit allowance on those utilities, hence more to sell. As your excess credits are sold, they can be applied to the cost of your wind- and solar-powered home and plug-in vehicle. Corporations in effect, will be purchasing carbon from citizens in the US and not Brazil or other less polluted countries. What do you think?

"Be the Change That You Want To See in the World"

Gandhi

Effort is Free!

! @ #. $ % ^ &, *" ?/>|?` ~ () + CO_2 ": Here are extra symbols and punctuations. If I have misplaced a few along the way, you can put these where you like.

~GREEN BEINGS —SAVE GENES~

Resource Guides A-Z
Informational Purposes Only

Author's Favorites
Must See Sights

The Dangers of Plastic Bags
http://lee.ifas.ufl.edu/FYN/FYNPubs/TheDangersofPlasticBags.pdf
The History of Stuff by Annie Leonardhttp://www.storyofstuff.com/
The Poison Plastic featuring Sam Suds by Free Range Studios
http://www.youtube.com/watch?v=qpmE_b90XTU
Global Warming. Global Action. Global Future.
http://www.350.org/#tabs-rotator-1
Mike Tidwell, Director, Chesapeake Climate Action Network, *When Words Fail* www.orionmagazine.org

Efficiency

http://www.aboutmyplanet.com/daily-green-tips/recycle-your-ink-cartridge#more-883
www.asa.org/section/_audience/consumers/
www.aceee.org

www.aceee.org/consumerguide/mostenef.htm
http://www.ase.org/
http://www.ase.org/content/article/detail/2654#credit Table
http://www.carbonrally.com/challenges
http://earth911.org/energy/energy-conversation-tips/
http://www1.eere.energy.gov/consumer/tips/home_energy.html
http://www1.eere.energy.gov/consumer/tips
http://www.energy.maryland.gov
http://energysavingnow.com
Energy Savers: http://www.eere.energy.gov/consumer/
ENERGY STAR www.energystar.gov
http://www.energystar.gov/ia/partners/promotions/change_light/downloads/Fact_Sheet_Mercury.pdf
http://www.fightglobalwarming.com/documents/5120_BrochureR4.pdf
http://www.greencommunitiesonline.org/about-essentials-grants.asp
http://www.grist.org/
Maryland Office of the People's Counsel
http://www.opc.state.md.us/
http://www.newdream.org/consumer/energy_checklist.html, http://www.newdream.org/buy/save_energy.php
http://www.newdream.org/consumer/doe_tips.html
http://www.newdream.org/consumer/rmi_energy_tips.html
http://www.newdream.org/consumer/tuneup.html
http://www.newdream.org/consumer/consumer_energy_tips.html

www.rmi.org
http://www.newdream.org/consumer/PKG22.pdf
http://www.nrdc.org/greenliving/toolkit.asp
http://www.opc.state.md.us/assets/documents/conservation_tips.pdf
http://www.psc.state.md.us/psc/index.htm
http://www.repp.org/efficiency/index.html
http://www.ucsUS.org/assets/documents/clean_vehicles/Vanguard-Technical-Report-final.pdf
http://www.ucsUS.org/clean_vehicles/vehicles_health/ucs-vanguard.html
http://www.wwf.org.uk/oneplanet/youcan_0000003955.asp
www.ase.org

LEED

Alliance for Green Development
Albuquerque, NM
(505) 269-2969
http://www.greenalliancenm.org/

Arlington County Green Home Choice Program
Arlington, VA
(703) 228-4792
http://www.arlingtonva.us/
Build San Antonio Green
San Antonio, TX 210-224-7278
http://www.buildsagreen.org/
Green Points Program
Green Built, Inc.
Grand Rapids, MI

(616) 281-2021

Green Home, Inc.
Washington, DC
(202) 544-5356
http://www.greenhome.org/

Green Homes Northeast
Boston, MA
(617) 374-3740
www.ghne.org
Green Home Program
New York City, NY
Building America
Nationwide
(202) 586-9472
www.eere.energy.gov/buildings/building_america

Solar

Energy-Efficient Home Plans for Kentucky, and Kentucky Solar Design Awards
Division of Energy
663 Teton Trail
Frankfort, KY 40601
(502) 564-7192

Home Designs for Energy Efficient Living
Home Styles Publishing and Marketing
275 Market Street, Suite 521

Minneapolis, MN 55405
(612) 338-8155
Operation Solar
Northeast Utilities
P.O. Box 270
Hartford, CT 06141-0270
(860) 665-5000

Partnership for Advancing Technology in Housing
http://www.pathnet.org

Plan book for Low-Cost Energy Efficient Homes
South-face Energy Institute
P.O. Box 5506
Atlanta, GA 30307
(404) 515-7657

Residential Solar Architecture: A Representational Inventory
Nebraska Library Commission
Interlibrary Loan
1200 N Street, Suite 120
Lincoln, NE 68508
(402) 471-2045

Solar Homes for North Carolina
North Carolina Solar Center
Box 7401 North Carolina State University
Raleigh, NC 27695-7401

(919) 515-3480
in NC, (800) 33-NC SUN
http://www.ncsc.ncsu.edu
Solar Homes for South Carolina
S.C. Energy Office
1201 Main Street, Suite 820
Columbia, SC 29201-3227
(803) 737-8030

Sun-Inspired Home Plans
Energetic Design
18250 Tanner Road
Citronelle, AL 36522
(334) 866-2574 suninspired@earthlink.net Debra Rucker Coleman,
Architect http://energeticdesign.homestead.com

Sunterra Homes
www.homesbysunterra.com
The New Florida Home,
The Florida Cracker-Style, and
The Expandable Affordable Home
Public Information Office
Florida Solar Energy Center
1679 Clearlake Road
Cocoa, FL 32922
http://www.fsec.ucf.edu/en

www.nrel.gov/rredc/wind_resource.html
www.eere.energy.gov/

www.findsolar.com

Other Books

Clark, Sarah L., *Fight Global Warming: 29 Things You Can Do.* (New York Consumer Report books in association with Environmental Defense Fund, 1991).

DeCicco, John, et al, *CO_2 Diet for a Greenhouse Planet: A Citizen's Guide for Slowing Global Warming*, (New York: National Audubon Society, 1990).

Rocky Mountain Institute. *The Efficient House Sourcebook and Home Made Money: How to Save Energy and Dollars in your Home.*
US DOE - Energy Efficiency and Renewable Energy Network (EREN) www.eren.doe.gov

Wilson, Alex, *1991 Consumer Guide to Home Energy Savings* (Washington, D.C.; American Council for an Energy-Efficient Economy, 1990).
Hybrid information tax credits
2005 Building Energy Data Book, Table 4.2.1 (PDF 132 KB). Download Adobe Reader.

Disclaimer Statement: The information in this book is believed to be current and correct; however, I do not guarantee that the information is current or correct. All information is readily available on the web and every attempt has been made to credit the resources as listed.

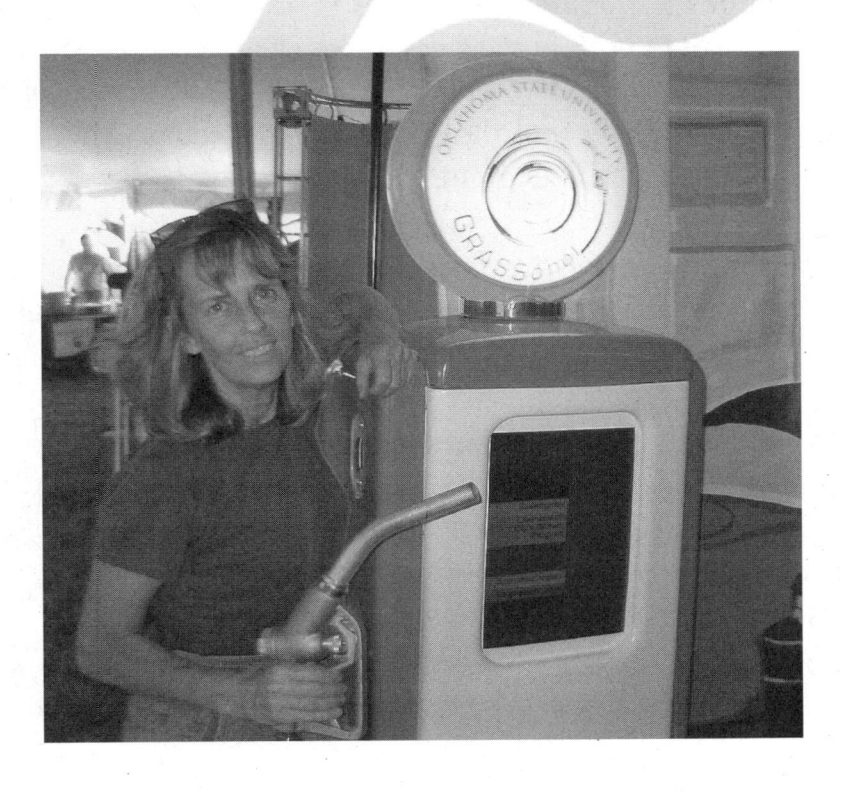

First in line at the Grassahol pump!

~GREEN BEINGS –SAVE GENES~

Glossary

Acid Rain is the result of sulfuric acid combining with moisture in the air such as rain or snow, which then becomes acidic. The acid damages plants, soil and coral reefs by lowering the ph of water. In time, acid rain lowers the ph of soil too: harming trees and many species including mankind.

Adjusted Fuel Utilization Efficiency (AFUE) is the amount of heat actually delivered to your house compared to the amount of fuel that you supply the furnace. Thus, a furnace that has an 80 percent AFUE rating converts 80 percent of the fuel that you supply to heat -- the other 20 percent is lost out of the chimney.

Air Pollution refers to three things: acid rain, greenhouse effect and ozone depletion. All are harmful to plants, animals and water on the earth.

Biofuels are renewable fuels made into liquid from the anaerobic digestion of plant matter or recently living organisms such as manure from cows. Being renewable, it is unlike petroleum or coal from fossil fuels, or nuclear fuel. An example of biofuel is recycled vegetable oil.

Biodiesel fuels are produced from any fat or vegetable oil through a refinery process. They contain no petroleum, are registered as a 100 percent alternative fuel; reduce net carbon dioxide emissions by 78 percent in comparison to petroleum diesel.

British thermal unit (BTU) is a grading standard that reflects the amount of heat required to raise the temperature of one pound of water one degree Fahrenheit. One Btu = 252 calories, 778 foot pounds, 1055 joules or 0.293 watt hours.

Carbon Dioxide (CO_2) is produced when fossil fuels are burned. Plants can convert carbon dioxide back to oxygen, but Earth's forests continue to be cut (less than four percent rainforest remain) and plants are damaged by acid rain, so the planet as fewer resources for absorbing the CO_2. This carbon acts like a blanket and traps heat in our atmosphere warming the planet. Scientists agree an increase in carbon correlates with the rise in global temperatures and recent climate changes. Although, many nations have been pledging steps to curb emissions for nearly a decade, the world's output of carbon from human activities is increasing and totals about ten billion tons a year and has been steadily rising.

Carbon-Neutral Design is a design process that doesn't produce greenhouse gases or contribute to global warming during any part of its life cycle, from production through consumption and disposal. By using daylight, ventilation, water use and other factors a building can be constructed without polluting the environment.

Coefficient Performance (COP) is a grading standard for gas furnaces. The rating scale is 0.78-0.94. The higher numbers reflect better the performance.

Confined Animal Feeding Operation is an animal "factory", a method of raising livestock that contributes the most significant and serious environmental problems on every measurable scale. See factory farm definition.

Conventional Power produced from non-renewable fuels, such as coal, oil, natural gas, and nuclear material is called conventional fuel. It is a finite resource that cannot be replenished once extracted and used.

Efficiency Factor (EF) is a grading standard for gas and electric water heaters, effective 2004. The standard grade should be 0.9 for a 50-gallon electric water heater and 0.59 for gas water heaters.

Eggs are classified as Certified Organic, Free Range, Cage Free, Natural, and Pasture Raised. Only a few of these classifications actually offer an opportunity of the hen to go outside for an hour or two each day, Cage free is restricted to a hen house. Pasture raised does not yet have a USDA certified label.

Energy Efficiency refers to products or systems that use less energy to do the same or better job than conventional products or systems. Energy efficiency saves energy, saves money on utility bills, and helps protect the environment by reducing the demand for electricity.

Energy Efficient Ratio (EER) is a grading standard for window or room air conditioners developed in October 2000. A rating of nine is low and eleven is considered excellent.

Energy Star is a Federal standard applied to equipment for the purpose of rating its energy efficiency.

Evacuated Tube Water Heaters is a method of heating propylene glycol and water solution by heating the tips of copper pipes. The long copper pipes are in an air-evacuated vacuum tube that is sealed. The liquid will boil in the presence of sunlight. This system even works well on cloudy days but must be free of snow and ice.

Factory Farming is an industrialized system of producing meat, eggs, and milk in large-scale facilities where the animal is treated as a machine. Some of the characteristics of a factory farm include intensive crowding of animals, trimming of birds' beaks, breeding and birthing cages for pigs' so small they cannot stand or turn over, force-feeding of ducks, injecting artificial growth hormones, antibiotics, steroids, restricting mobility, and animals can be fed rendered meat food even though they are herbivores. Also known as confined feeding operations (CAFOs).

Fossil Fuels supply the nation's principal sources of electricity. The popularity of these fuels-coal, oil, and natural gas-is largely due to their low costs. Because fossil fuels are a finite resource and cannot be replenished once they are extracted and burned, they are not considered renewable.

Geo-Exchange or Geo-Thermal Heat Pump systems use the Earth as a source of heat in the winter and a heat sink in the summer. It is rated with the COP scale. Geo-exchange averages between three and four.

Geo Thermal (Geo (Earth), Thermal (heat)) is clean renewable heat from the center of the Earth relies on endless supply of core heat expected to last for billions of years and is equal to 42 million megawatts of power. One megawatt, equivalent to one million watts, can meet the power needs of about 1,000 people. The geothermal resources used throughout the US today serve 2.8 million households.

Global Climate Change refers to any significant change in measures of climate such as temperature, precipitation, or wind lasting for an extended period of time (decades or longer). Climate change may result from natural factors such as the sun's intensity, changes from ocean circulation and human activities that change the atmosphere, land and water. See Carbon Dioxide.

Greenhouse Gases (GHG) refer to carbon dioxide, nitrous oxide and methane gas, halogenated fluorocarbons, ozone, perfluorinated carbons, and hydrofluorocarbons emissions that all cause global climate change. It is measured by metric tons of carbon equivalent (MTCE). These toxins are found in solid waste as well as automobile exhaust.

Green Homes use less energy, water, and natural resources, create less waste, and are healthier for the people living inside.

Green Power refers to renewable energy resources such as solar, wind, geothermal, biogas, biomass and low-impact hydro-generated green power. A green power resource produces electricity with zero anthropogenic (caused by humans) emissions, has a superior environmental profile to conventional power generation, and must have been built after the beginning of the voluntary market (1/1/1997).

Heating Seasonal Performance Factor (HSPF) is a grading standard for electric heat developed in January 2006. A rating of 7.7 is low and 10 are considered excellent.

Inverter is a device that converts direct current (DC) to alternating current (AC) electricity. The device synchronizes your electricity with that of the power company's grid electricity. It has an automated power-off switch to prevent power to the grid while the grid is down due to weather conditions etc. This protects the utility worker from electrocution.

Kilowatt Hour (kWh) means one thousand units. One kilowatt-hour is equal to 1,000 watts of electricity used continuously for one hour. The average home in the US uses approximately 877 kWh/month of electricity. To calculate the cost of any electrical devise, check the wattage x hours used/1000=kWh. Then kWh x cost per kWh=operating cost. One-kilowatt hour of electricity produces 1.4 pounds of carbon dioxide. Example: A 100watt light bulb in use for ten hours uses 1000 watt-hours, or one kilowatt of electricity. (100 watts x 10 hours = 1000 watt-hours = 1 kWh).

Megawatt (M) means one million watts. One megawatt-hour (mWh) is equal to 1,000 kilowatt-hours.

Modified Energy Factor (MEF) is a grading standard for clothes washers produced since January 2004. A low standard is 1.04 and a high-energy efficiency level is 1.26.

Net Metering is a method of crediting customers for electricity that they generate on-site in excess of their own electricity consumption. In an energy efficient home, DC power is sent to the grid. With an inverter, it is converted to AC power. Customers with their own generation of DC power offset the electricity they would have purchased from their utility. If such customers generate more than they use in a billing period, their electric meter turns backwards to indicate their net excess generation. Depending on individual state or utility rules, the net excess generation may be credited to their account (in many cases at the retail price), carried over to a future billing period, or ignored.

Ozone is a chemical compound made of three oxygen atoms created from a chemical reaction between heat and pollutants. Breathing ozone is harmful to health. When levels are high, a Code Red alert is announced to the public.

Passive Solar Heat provides cooling and heating to keep a home comfortable without the use of mechanical equipment. This style of construction results in homes that respond to the environment and amount of sun exposure in the house.

Photovoltaic Module is a device made from silicon, boron and phosphorous which creates an electrical current through exposure to sunlight. These devices are manufactured in various sizes of voltages and wattage.

R-Value is a grading standard that indicates insulation's resistance to heat flow. The higher R number in the grading in scale indicates a better quality of the insulation.

Seasonal Energy Efficient Ratio (SEER) is a grading standard for central air conditioners. A rate at the level of 13 is low and 15 are considered excellent.

Solar Heat Gain Coefficient (SHGC) is a standard used in passive solar design that demonstrates the solar heat transmitted directly through the window or door and absorbed into the home as heat. A lower SHGC rating provides less solar heat transmission and is effective in decreasing cooling needs. More than 0.6 SHGC is effective in collecting solar heat gain during the winter months.

Super-Ultra-Low-Emission Vehicles (SULEVs) is a special California regulatory designation greater than Partial Zero-Emission Vehicle (PZEVs) for vehicles having greater emissions control-system durability and emitting near-zero levels of evaporative hydrocarbons.

Sustainability is a term that relates to something that can continue indefinitely. To build a sustainable society, we must be careful about how we use energy to avoid ruining the environment for ourselves and other animals and plants.

Thermal Collectors and Solar Hot Water Heaters are solar collectors that circulate a propylene glycol and water solution through an exchanger that will heat your domestic hot water. This system requires that local plumbing codes be followed. Newer systems have been created. See Evacuated Tube Water Heaters.

U Factor rates an entire window's performance, including frame and spacer material, door, and also skylights conducting non-solar heat flow. The lower rating of the U-factor (0.35 or less) indicates the better energy-efficient the window, door, or skylight.

Visible Transmittance (VT) is an optical property that indicates the amount of light through the window. The scale is from 0-1. Most windows are in the range of 0.3-0.8. The higher the number, the more visible light transmitted through the window.

Volatile Organic Compounds (VOCs) are toxic chemical compounds that have high enough vapor pressure under normal conditions and room temperature to vaporize and enter the atmosphere—a process called "off-gassing." Common VOCs such as formaldehyde, acetone, benzene, vinyl, some plastics and gasoline can be found in paint, paint thinners, carpeting, furniture, flooring and household cleaning products. .

Watt is a scale to measure amount of energy used for electric power. One watt is the amount of power rate of one joule of work per second of time. For example a 100 watt light bulb uses 100 watts of energy instantly as it is turned on. If it remains on for three-hours (3hrs bx100watts) the light has uses 300 watts of power.

Index

aerators, 144

ALTO® Lamp, 146

Attic fans, 172

Awnings, 168

baking soda, 174, 177

bamboo flooring, 198, 210

bamboo towels, 161

bamboo) towel, 35

bat box, 78

bees wax candle, 98

bicycle, 256

Bio-bags, 80

Borax, 174, 176

Boric acid, 178

brick, 13, 94

bubble wrap, 81, 117

can opener, 146

cards, 71, 120

caulking, 178

Caulking, 178

caulking gun, 179

Cellular shades, 165

Christmas lights, 153

cloth napkins, 114

clothes rack, 91

clothesline, 91

coffee cup, 100

coffee filter, 148

Compact fluorescent light, 154

Compost Bin, 164

compost crock, 80

compostable utensils, 147

composting toilet, 13

computer monitor, 168

cork bottle stoppers, 142

cork screw, 142

cutlery, 149

diapers, 160

Diapers, 160

digital thermometer, 148

digital-to-analog converter, 109

Drainbo, 150

Duraflame, 131

Fair-Trade, 86, 297

Fans, 162, 172

filter whistle, 103

fluorescent bulbs, 31, 153, 156

furnace filter, 102

garment, 106, 107, 149

garment bag, 106

gaskets, 35

Global Basket, 124

global positioning system, 58

grocery bag, 148

houseplant, 83

Insulation, 140, 199

kilowatt meter, 159

laundry soap, 93, 176

LED holiday, 153

light detectors, 147

light sensor, 147

Melaleuca, 175

mower, 81, 82, 263

nature mill, 80

Nature Mill, 171, 172

nozzle, 65, 158

organic wine, 85

paint, 101, 102, 171, 293

Paint, 101, 170

phosphate free, 92, 93, 104

Piggy Bank, 138

pots, 84, 177

power monitor, 167

~GREEN BEINGS –SAVE GENES~

power strip, 157

programmable thermostat, 70, 166

rain barrel, 79, 163

rechargeable batteries, 115, 146

Rechargeable batteries, 152

Recycled paper, 144

Resource Guides, 276

reusable bag, 60, 122

reusable bags, 121, 122

reusable grocery bags, 121

reusable water bottle, 96

shades and blinds, 47

shower timer, 39

showerheads, 150, 151

Showerheads, 151

sign, 47, 60, 267

Sink Positive, 166

Solar blankets, 169

Solar charger, 163

solar chargers, 153, 166

solar landscape lights, 152

solar oven, 170

Solar tubes, 169

tinting, 61, 197

tire gauge, 56

tooth-brush, 125

trees, 46, 62, 70, 76, 89, 90, 108, 114, 117, 122, 143, 144, 148, 151, 199, 210, 243, 262, 264, 284

trekking sticks, 256

video, 82, 99, 111, 116

videos, 117

Videos, 50

Vinegar, 177

water heater blanket, 157

weighted with pebbles, 94

A GLOBAL BASKET, What's in it for me?

Take a step further: The latest research claims that *happiness can be purchased*. But it seems it will only be your happiness if you give it away. Experimental subjects have reported significantly greater happiness if they spend money on a gift rather than themselves. And here is your opportunity to receive a lot of happiness by giving a unique basket of environmental products along with this book. This is the most worthwhile gift of care and concern you can give. Happiness and a warm-hearted feeling are guaranteed.
http://www.news.harvard.edu/gazette/2008/04.17/31-happiness.html

Soon, with the index of this book you too may select and design your own basket of energy filled products. This distinctive Eco-Starter Kit is presented, packaged and delivered in a Ghana Fair-Trade handmade basket. Use the basket as your designated receptacle for all those recyclable products we dispose of in the kitchen. For example: glass, cans, paper, plastics and junk mail can be beautifully concealed in the basket. Or carry it to the farmers market for your local produce. Onlookers will stop you just to inquire the where-abouts of such a unique basket. This gives you the opportunity to share your knowledge about energy efficiency, and the chain link will just keep growing.

Women gather in a co-op, thousands of miles away every day to weave these baskets designed for your energy saving efforts in hopes of making the world a little greener for us all. It takes one complete day to weave the elephant grass into the tightly woven basket. Each one is different. Plant dyes are used for intricate designs and colors depending on availability. This makes an exceptional gift for everyone, including you. Send a little happiness to someone special.

Please Visit

GlobalBasketcase.com

This book has been printed on 100 percent recycled paper. The blue soy-based ink to remind you this book is your 'blue print' for energy efficiency to keep as a reference and as a guide to a healthier and richer way of life. And the price, 1350 remember, 1 earth = **350** ppm.

If I have overlooked your efficiency tactics or clever idea, please email me with your thoughts for the next edition.
suz@greenbeanpublishing.com

Thank you for **WALKING THE TALK** of solutions toward Energy Efficiency.

And remember to smile, It uses less energy!

Recycled Paper FYI

The US is the largest consumer of paper and paper products in the world, using 90 million tons of paper annually. Today more than 90 percent of paper comes from tree pulpwood. **Ninety-five percent of new books come from sacrificed trees!**
  http://www.fscus.org/paper/

The average tree is about 680 kg, or 0.68 of a ton. It is estimated 16.32 million trees are harvested for paper production in the US each year.
http://www.triplepundit.com/pages/askpablo-how-many-trees-does-m-002613.php
But there is a way to stop this destruction; **Recycled paper!**

Recycled paper is available in three categories. Pre-consumer waste, post-consumer waste or mill broke (paper trimmings or scrap paper from the production of paper in mills) **Post-consumer waste** or **PCW** means that the product's pulp was made from paper once used by consumers; **100 percent PCW is most desirable.** No independent certifier regulates it. But, the Federal Trade Commission has defined its use and the terms.
www.ftc.gov/bcp/grnrule/guides92.htm#G5
- Recycled paper requires 45-65% less energy in production
- 36% less water pollution, by using fewer chemicals
- 74% less air pollution compared to making new paper.

Have you ever driven past a paper factory? PEEEU!

Another alternative is Forest Stewardship Council FSC-certified paper. Their logo tells you the manufacturers support the highest social and environmental standards in the market for new paper or other products manufactured from wood.

Some companies use 10-30% post consumer paper and mix it with Forest Certified new paper.

Tree-free means that the paper was made from kenaf, hemp, bamboo, agricultural refuse old blue jeans or other non-tree fibers. The use of this term is currently unregulated.

The good news is you don't need to go to China to find recycled paper anymore! With many thanks and sincere appreciation for the insight of this business owner, this book was printed on 100% PCW. Thank you to all United Book Press employees!

United Book Press
1807 Whitehead Road
Baltimore, MD 21207
(Ph) 410.944.4044
(Fx) 410.944.4049
email: sbupp@unitedbookpress.com

Notes

Notes

Notes